An Introduction to

ASTRONOMY AND ASTROPHYSICS

An Introduction to
ASTRONOMY AND
ASTROPHYSICS

Pankaj Jain

Indian Institute of Technology
Kanpur, India

CRC Press
Taylor & Francis Group
Boca Raton London New York

CRC Press is an imprint of the
Taylor & Francis Group, an **informa** business

A CHAPMAN & HALL BOOK

CRC Press
Taylor & Francis Group
6000 Broken Sound Parkway NW, Suite 300
Boca Raton, FL 33487-2742

© 2015 by Taylor & Francis Group, LLC
CRC Press is an imprint of Taylor & Francis Group, an Informa business

No claim to original U.S. Government works

Printed on acid-free paper
Version Date: 20150203

International Standard Book Number-13: 978-1-4398-8590-1 (Hardback)

Visit the Taylor & Francis Web site at
http://www.taylorandfrancis.com

and the CRC Press Web site at
http://www.crcpress.com

This book is dedicated to the memory of my father, Paras Prabhu Jain.

Contents

Preface

This book arose out of an introductory course in astronomy and astrophysics that I have offered several times at the Indian Institute of Technology, Kanpur. It is aimed at sophomore or junior level students, assuming that they have done a basic one-year coursework in mathematics and physics. I personally find astronomy to be a very interesting subject and hope that I have been able to share some of this enthusiasm in my book. Everyone is fascinated and intrigued by the heavens and look forward to an opportunity to get a deeper understanding about their nature. It is absolutely remarkable that we can comprehend the heavens in terms of the basic laws of physics. Hence besides providing a basic description of the heavenly objects, an introductory book on this subject must also relate their properties to the fundamental laws of physics. This makes the subject somewhat challenging. At the same time one gets a much better appreciation of these laws by learning their direct application to astronomical objects. In this book I have tried to explain the application of physical laws to astronomy and astrophysics to the best of my abilities. In many cases these applications are quite subtle and may require a careful reading before they can be appreciated. I have also briefly introduced basic concepts in physics and mathematics wherever they are required.

The book is relatively concise and does not attempt to cover all possible topics. I have tried to focus on basic ideas and have provided sufficient details so that the reader can appreciate the subject and be ready for advanced coursework. I have gone into considerable detail for some chosen topics which I thought were important and could be explained at this level. In other cases I have only provided a glimpse into the basic ideas and the reader is expected to consult other sources for a proper understanding.

The book can be used for a one- or a two-semester course in astronomy and astrophysics. It is not possible to cover all the topics in one semester. Hence if it is used for a one-semester course, some of the topics have to be skipped. In order to save time, it is possible to omit some mathematical derivations without distorting the logical flow of the subject. My experience has shown that students find cosmology particularly interesting and hence it should preferably be included. In any case, I feel that a proper introduction to the subject requires two semesters of coursework. If possible, the course should also provide some hands-on experience with observing the night sky and analyzing astronomical data. For this purpose several excellent web resources are available.

I acknowledge considerable help from Roshan Bhaskaran in preparing figures for the book. I have also made extensive use of the software Stellarium in some of the chapters and thank its developers for producing such a useful software. Finally, I would like to thank the scientific staff members of organizations, such as NASA; European Southern Observatory; Institute for Solar Physics, Stockholm, Sweden; High Altitude Observatory; Yohkoh Legacy Archive; and the National Radio Astronomy Observatory for producing fascinating astrophysical images and making them available publically.

Chapter 1

Introduction

1.1   Overview

The Universe is fascinating. Starting in ancient times, people have wondered about the nature of stars and how they might affect our lives. Our knowledge about them has progressed with time, often assisted by advances in fundamental science. For example, in the nineteenth century, the source of solar energy was believed to be the gravitational potential energy and perhaps the chemical energy. These were the only possibilities known at that time. It was only later, after the discovery of nuclear fusion, that scientists realized that the Sun is powered by nuclear energy. At present we are exploring the Universe in great detail with very sophisticated instruments. This has led to a huge amount of very precise data on various astronomical structures. The data has provided us with in-depth knowledge and understanding of the Universe on different length scales. Hence we are now able to construct detailed models, assisted by computer simulations, of objects such as stars, stellar clusters, galaxies etc. The branch of science that aims to understand the physics and chemistry of such heavenly objects is called Astrophysics. This field has seen remarkable developments in the last century and has now reached the level of a precision science.

The Universe contains structures on a wide range of scales. These include the solar system and the planetary systems associated with other stars. The stars themselves often form clusters that are part of bigger structures called galaxies. Furthermore, the galaxies are also not found in isolation and form groups or clusters of galaxies that form larger clusters called superclusters. The superclusters are the largest structures observed. The size of the observable Universe is roughly 50 times larger than the size of the largest supercluster.

The Universe is continuously evolving. The planets and the stars, which appear to us as everlasting, originated at some time in the past. The stars evolve with time, slowly exhaust their nuclear fuel, and eventually die. Similarly, the galaxies also have an origin and slowly evolve into the structures we observe today. It is conceivable that the entire Universe also had an origin. The model that appears to best describe the origin and evolution of the Universe is called the Big Bang model. This model postulates that the Universe was initially very small in size. It was also very hot and dense and consisted predominantly of radiation and plasma, that is, charged particles such as electrons and protons. With time it expanded, cooled, and its density decreased. At some stage the Universe was cool enough to allow formation of the atomic nuclei of a few light elements. At this time the Universe was opaque to photons, which could travel only a very short distance before being scattered or absorbed by an electron or a proton. As it cooled further, neutral atoms began to form. The photons do not interact as strongly with atoms as they do with free electrons and ions. Hence around that time, the photons could travel freely in the Universe. The Universe became transparent and electromagnetic radiation, such as visible light, was able to propagate large distances through space. As it cooled further, structures such as galaxies, clusters of galaxies, and stars started forming due to gravitational attraction. The Universe slowly evolved into the present state observable today.

1.2   Scales and Dimensions

We next give a brief overview of the scales and dimensions of various objects and structures in the Universe. Let's start with the Earth. The radius, R_E, and mass, M_E, of the Earth are

$$R_E = 6378 \text{ Km},$$

$$M_E = 5.974 \times 10^{24} \text{ Kg}.$$

In comparison, the Sun is about a million times more massive with about 100 times larger radius:

$$R_S = 6.96 \times 10^5 \text{ Km},$$

$$M_S = 1.989 \times 10^{30} \text{ Kg}.$$

The Earth-Sun distance is called one Astronomical Unit (AU),

$$1 \text{ AU} = 1.496 \times 10^8 \text{ Km}.$$

The Sun is the nearest star. The next nearest star is Proxima Centauri, at a distance of 1.31 pc. Here, pc denotes parsec and is given by

$$1 \text{ pc} \approx 3 \times 10^{13} \text{ Km}.$$

Hence, the next nearest star is roughly 200,000 times the distance to the Sun.

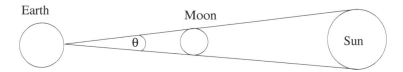

FIGURE 1.1: The angle, θ, subtended by the Sun at Earth is approximately the same as that subtended by the Moon on Earth.

Using the information given above, we can compute the angle θ subtended by the Sun's diameter at Earth. We find

$$\theta = \frac{2 \times 6.96 \times 10^6}{1.496 \times 10^8} = 0.0093 \text{ radians}.$$

This is roughly equal to 0.5 degrees. Coincidently, this also happens to be roughly the angle subtended by the Moon on Earth (see Figure 1.1), although its distance and size is very different from that of the Sun. If the angle subtended by the Moon had been much smaller, we would not have been able to observe the spectacular phenomena of solar eclipses.

As we have already mentioned, stars are not uniformly distributed in the sky but are found to occur in clusters called galaxies. Galaxies are organized structures, with a high density of matter near the center. Many galaxies show evidence of rotation about the center. A typical galaxy contains billions of stars. For example, we are part of the Milky Way galaxy, which consists of over 200 billion stars. The Milky Way is a spiral galaxy and forms a disk-like structure with a disk diameter of about 30 Kpc (1 Kpc = 1,000 pc) and disk thickness of 1 Kpc (see Figure 1.2). Near the center of the galaxy the density of matter and stars is very high. Here the Milky Way galaxy displays a ellipsoidal bulge-like structure with the major axis equal to a few Kpc. The Sun is located at a distance of about 8 Kpc from the center of the galaxy. The Milky Way is visible as a white band stretching across the night sky, provided one does not have too much city light pollution. Because we are located inside the Milky Way, we cannot see its spiral structure. In order to observe the spiral arms, we require a view from above or below the disk. Surrounding the disk is a nearly spherically symmetric structure called the galactic halo. Here the density of matter is much lower than the disk.

The observable Universe contains on the order of 100 billion galaxies. The typical distance scale of galaxy clusters ranges from 1 to 10 Mpc (1 Mpc = 1 million pc) and that of superclusters is about 10 to 100 Mpc. Superclusters are the largest structures seen in the Universe. If we take a broad-brush view of the Universe at distance scale much larger than 100 Mpc, the Universe appears to be uniform, that is, on average the distribution of matter is the same in all directions and positions. The size of the observable Universe is on the order of a few Gpc (1 Gpc = 10^9 pc).

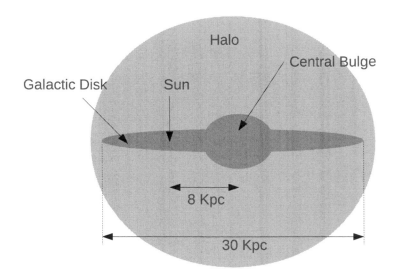

FIGURE 1.2: A schematic illustration of the Milky Way. The diameter of the galactic disk is about 30 Kpc while its thickness is about 1 Kpc. The Sun is located roughly 8 Kpc from the center of the galaxy. The halo extends to much larger distances beyond the disk.

Let us recall that light travels at a finite speed, which is equal to $c = 2.998 \times 10^8$ m/s. It takes light about 8 minutes to reach us from the Sun. In general, the light that we receive from an object at distance D left the object at time D/c ago. The next nearest star is located at a distance of 1.31 pc. Besides parsec, another convenient unit of distance in astronomy is a light year. This is defined as the distance that light travels in 1 year. The relationship between parsec and light year is given by

$$1 \text{ pc} = 3.26 \text{ light years}. \tag{1.1}$$

Hence the next nearest star is 4.3 light years away. This means that light takes 4.3 years to reach us from this star. Similarly, light we receive from the Andromeda galaxy, located at a distance of about 1 Mpc, left this galaxy about 3 million years ago. The current best estimate of the age of the Universe is about 14 Giga years (1 Giga year $= 10^9$ years). Hence the farthest we can observe today is the distance that light travels in this time. This distance is equal to 14 Giga light years, which is approximately equal to 4 Gpc. This provides us with an estimate of the size of the observable Universe.

The Universe changes slowly with time. All structures in the Universe, such as stars, galaxies, galaxy clusters, undergo evolution. For most astronomical objects this evolution is very slow and we are unable to directly observe it. However this evolution can be observed indirectly. For example, we observe different types of stars, such as main sequence stars, red giants, white dwarfs etc. By the theory of stellar evolution, we deduce that these represent different stages of evolution of stars. Hence by observing a red giant star of mass approximately equal to 1 solar mass (M_S), we can see how the Sun, which is currently a main sequence star, might appear during the late stages of its life cycle. Furthermore, due to the finite speed of light, as we see objects that are very far away from us, we see them at much earlier times. The far away objects, therefore, give us a glimpse of the Universe in its early stage of evolution. For example, as we observe objects at a distance of 1 Gpc, we are effectively observing how the Universe appeared 3.26 billion years ago.

1.3 Night Sky

In order to gain a proper appreciation of astronomy, a person should become familiar with the night sky. Unfortunately with the high pollution levels and lights in cities, the number of stars visible is relatively small. In order to get a better view, it is useful to go to the countryside.

The night sky is continuously changing due to the rotation of the Earth about its axis. The stars appear to move from east to west. This apparent motion of stars is due to rotation of the Earth. The stars themselves remain approximately fixed with respect to the Sun. The Earth rotates about an axis, which is perpendicular to the equatorial plane, as shown in Figures 1.3 and 1.4. The equatorial plane separates the northern and southern hemispheres and passes through the Earth's equator. The sense of rotation of Earth is from west to east. Due to this rotation, the Sun and all the stars appear to rise from the east and set in the west. Furthermore, the night sky also changes due to the revolution of the Earth around the Sun. The plane of revolution around the Sun is inclined at an angle of about 23.5° with respect to the equatorial plane. This plane is called the ecliptic plane, as shown in Figure 1.5. The sense of revolution of the Earth is also from west to east.

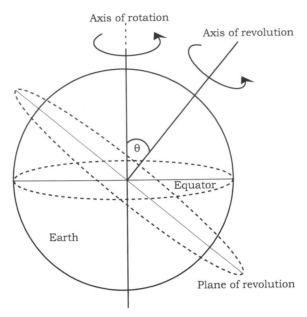

FIGURE 1.3: The axis of rotation of the Earth is perpendicular to the equatorial plane. The plane of revolution of the Earth around the Sun is inclined at an angle $\theta = 23.5°$ with respect to the equatorial plane.

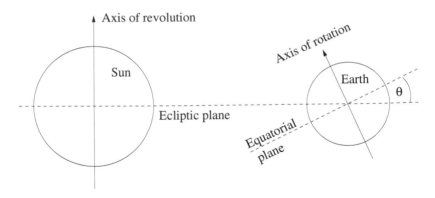

FIGURE 1.4: The plane of revolution of the Earth around the Sun is called the ecliptic plane.

The Sun lies in different directions with respect to the Earth at different times of the year, as shown in Figure 1.5. Hence during night time we get a different view of the sky on different dates. For example, in the northern hemisphere, some of the prominent stars in summer are Vega, Altair, and Deneb, whereas in winter the prominent stars are Capella, Betelgeuse, Sirius, etc.

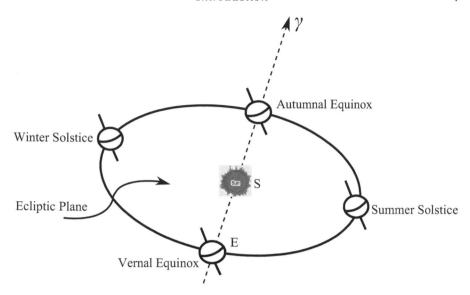

FIGURE 1.5: The orbit of revolution of the Earth around the Sun lies in a plane, called the ecliptic plane. The position of Earth at four different times is shown. The axis of rotation of Earth as well as the equatorial plane are also shown in the figure. The Sun as viewed from Earth at the time of vernal equinox lies in a direction in space, also called the vernal equinox.

As we view the night sky we can see a large variety of objects such as stars, planets, shooting stars or meteors, meteorites, comets, or an artificial satellite. A star is approximately fixed in space. It appears to move only due to the motion of the observer on Earth. However, its position relative to other stars or with respect to the Sun does not show much change over a period of days or even a year. It changes at a very slow rate due to the proper motion of stars. A planet moves around the Sun in an elliptical orbit. Hence its position changes significantly with respect to the background fixed stars.

There also exist many interesting phenomena caused by smaller objects in the solar system called meteoroids. These objects may sometimes enter the Earth's atmosphere at high speeds on the order of 10 to 75 Km/s. Due to friction, they heat up and become illuminated at an altitude of about 100 Km. A small meteoroid lasts a very short time interval. It produces a short streak of light that is called a meteor or a shooting star. The smallest meteoroid that can lead to a meteor observable by the naked eye has a mass of about 1 gram. Normally one sees some isolated meteors. On some dates one might see a meteor shower, containing a large number of meteors, which appear to diverge from a small region called the radiant. This happens when the Earth crosses some dusty regions in the solar system and hence undergoes collision with a large number of meteoroids. If the mass of the meteoroid exceeds 1 Kg, then it might survive until it reaches ground level. The resulting fragment is called a meteorite.

Another fascinating object that might be seen in the night sky is a comet. However, most of these are too dim to be seen by the naked eye. On average, one might be bright enough to be observable without a telescope every 10 years. They consist of a nucleus and a tail. The nuclei are composed of ice, rock, and dust and have diameters ranging from a few hundred meters to tens of kilometers. Comets move in a highly eccentric orbit around the Sun and are illuminated by sunlight.

1.4 Constellations

The stars are grouped into different constellations. Historically, these groupings were made on the basis of some figure these groups of stars appear to resemble. The choice of these figures was made entirely on the basis of human imagination. Constellations played an important role in ancient times because they were very useful for navigation. A person trained in reading the night sky could use them to identify different directions.

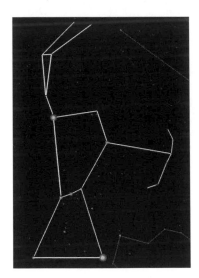

FIGURE 1.6: The Orion constellation, named after the mythological hunter, Orion. The brightest stars in this constellation are Rigel and Betelgeuse.

There are a large number of mythological stories associated with stars and constellations. Most civilizations have their own stories. Historically, they probably played an important role because they provided mnemonic devices to help recognize the different constellations. They also make the night sky become alive. Instead of just viewing stars, one now sees gods, goddesses, saints, warriors, nymphs, etc. in the night sky. It is filled with drama, action, and poetry.

Here we relate one story from Greek mythology. In the star cluster, Plaeides, the seven stars were originally seven sisters, daughters of the god Atlas and goddess Pleione. Once they were enjoying a walk in a forest with their friend, Artemis, the goddess of the Moon. Orion, the great hunter, happened to notice them and started to chase them. In order to save the seven sisters, Artemis asked the god Zeus for help. Zeus turned all of them into doves so that they could fly away. The sisters flew very high up in the sky and eventually became the Plaeides star cluster. This did solve the problem but was not quite what Artemis wanted. She became very unhappy because she had lost her good friends. Since Orion was responsible for all this, she decided to get even with him. With the help of her brother Apollo, she had Orion killed by a giant scorpion. Zeus then decided to put Orion up in the heavens by the side of the Plaeides sisters. As the stars move from east to west, Orion still appears to be chasing the sisters in the night sky. Furthermore Artemis, being the goddess of the Moon, gets to visit her friends sometimes in the sky. Finally, the giant scorpion became the constellation Scorpio. Scorpio rises from the east as soon as Orion sets in the west. So they are kept as far apart as possible.

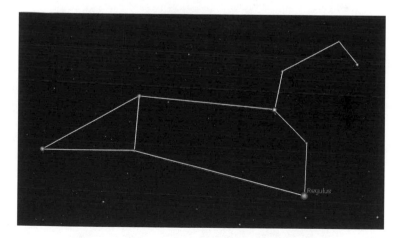

FIGURE 1.7: The Leo constellation which roughly appears as a lion. The brightest stars in this constellation are Regulus and Beta Leonis.

The constellation boundaries were made precise by the International Astronomical Union (IAU) in 1928. The entire sky was divided into 88 constellations. The boundaries were chosen on the basis of constant right ascension and declination using the equatorial coordinate system based on epoch B1875. The precise definition of this system is given in chapter 3. The coordinate system is updated every 50 years in order to account for the precession of the Earth's rotation axis. Hence, these boundaries do not correspond to constant right ascension and declination in the equatorial system, J2000, currently in use. These constellations play an extremely important role if one wants to recognize a particular star in the sky. A person trained in recognizing the constellations can identify a star even if he/she does not know its precise coordinates. Some examples of the constellations are Andromeda, Aries, Centaurus, Cygnus, Orion, etc.

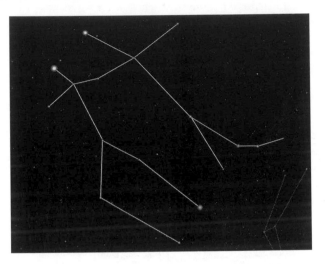

FIGURE 1.8: The Gemini constellation takes the shape of twins standing next to one another. Gemini is the Latin word for twins. The two brightest stars in this constellation are named after the twins, Pollux and Castor, in Greek mythology.

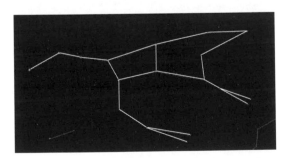

FIGURE 1.9: The Ursa Major constellation is also called the Great Bear. It contains seven bright stars seen in the upper left region of this figure. These are easily identifiable in the night sky and are collectively called the Big Dipper. These stars are also helpful in locating the North Star.

In Figures 1.6 through 1.9, we show some of the prominent constellations visible in the Northern hemisphere at different times of the year. We encourage the readers to view the night sky in order to identify some prominent constellations at their location. The stars in each constellation are labeled by the Greek letters α, β, γ depending on their brightness. For example, Vega, the brightest star in constellation Lyra, is denoted α Lyr. However, there are many exceptions to this rule, perhaps because the data on some of the stars were not available precisely at the time of this nomenclature. Hence in many cases, a star labeled as α is not the brightest star in the corresponding constellation.

1.5   Earth, Sun, and the Solar System

The Sun and all the planets in the solar system are shown schematically in Figure 1.10 in the order of their distance from the Sun. Hence the closest planet from the Sun is Mercury and the farthest is Neptune. Mercury is also the smallest planet while Jupiter is the largest.

All the planets revolve around the Sun in approximately circular orbits. The orbits of all the other planets also lie very close to the ecliptic plane. Hence the entire solar system forms a nearly planar structure, with the Sun at its center. The sense of revolution of all the planets is also the same as that of Earth. They all revolve from west to east. The trajectory of a planet appears very simple if viewed by an observer located at the Sun. However, it appears complicated if viewed from Earth.

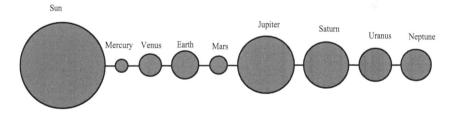

FIGURE 1.10: The Sun along with the eight planets. The distance and size is not shown to scale. However, the true order of the size and distance from the Sun corresponds to the order shown in figure.

As discussed earlier, the equatorial plane of Earth is inclined at an angle of about 23.5° to the ecliptic plane, as shown in Figure 1.4. The seasons are caused by this inclination. On March 21, the Sun's rays are directly overhead at the equator. This event and the corresponding position of the Sun is called spring or vernal equinox, as shown in Figure 1.5. During summer months the Sun is overhead at the northern latitudes. On June 21, it is overhead at the Tropic of Cancer, which is 23.5° north. This is the maximum northward motion of the Sun. On September 22, the Sun's rays are again overhead at the equator. Eventually on December 21, it reaches farthest south.

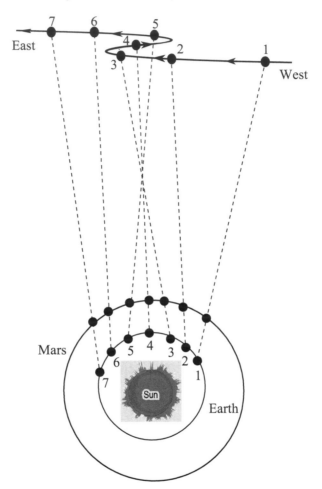

FIGURE 1.11: The retrograde motion of Mars, as seen from Earth, with respect to the fixed background stars. Mars normally appears to move from west to east. However, when it approaches Earth, it appears to move backward, makes a loop, and then again resumes its normal motion toward the east.

1.5.1   Retrograde Motion of Planets

Let's consider the motion of a planet relative to the fixed background stars. This is convenient because it eliminates the effect of the rotation of Earth.

Relative to the background stars, the planets appear to move in one direction and then may change direction for some time and again start moving in their original direction. Hence, for some time, they appear to move in a loop, as shown in Figure 1.11. This is called the retrograde loop.

Historically it was believed that the Earth is the center of the Universe, and that the Sun, stars, and all the other planets revolve around the Earth at uniform rates. The Earth was assumed to be stationary. This is called the geocentric theory of the Universe. The motion of the planets did not agree with the principles of this theory. They did not show uniform motion. With respect to the background stars, they showed retrograde loops, which were very puzzling from this point of view. However, Copernicus argued that these retrograde loops were explained nicely by the heliocentric theory, which proposes that all planets, including the Earth, revolve around the Sun. Furthermore, with the advent of telescopes, Galileo could observe the moons of Jupiter. This provided clear evidence of the existence of heavenly bodies that revolve around centers different from Earth. Hence, Earth no longer appeared to possess a special place in the Universe and this made it easier for the community to accept the heliocentric theory.

In Figure 1.11 we explain how the heliocentric theory explains the retrograde loop of Mars. As shown in Figure 1.11, Mars is further away from Sun in comparison to Earth. Its angular speed of revolution around the Sun is smaller in comparison to Earth. Hence it takes longer to complete one revolution. The position of Earth and Mars at several different times are shown in Figure 1.11. It is clear that for an observer at Earth, Mars would appear to traverse a retrograde loop.

1.6 Sidereal Time

The standard solar time is defined with reference to the Sun. One solar day is defined as the time interval after which the Sun is again at maximum elevation in the sky. For astronomical purposes, it is convenient to define sidereal time, which uses the fixed stars as a reference. Consider an observer P on the surface of Earth at 12 noon when the Sun is directly overhead on day 1, as shown in Figure 1.12. At this time, both the Sun and a fixed star are directly above P. In Figure 1.12, it is day time for P and hence the star will not actually be visible. After 1 sidereal day, P is again directly below the fixed star. The length of this day is a little shorter than a solar day because Earth has to rotate an extra angle, θ, before the Sun is again directly above P. On day 1, both the Sun and the star are overhead at noon. The next day, the Sun will again be overhead at noon, but the star a little earlier by an angle θ, converted to time units ($360° = 24$ hours). After 6 months, Earth is on the opposite side of the Sun. Now the star is overhead P at 12 midnight. The Sun comes overhead half a day later, that is, at noon. This implies that the number of sidereal days that have passed by now are half a day more than the solar days. Hence, in 1 year, the number of sidereal days completed is one more than the number of solar days.

Let T and T_1 represent the time intervals corresponding to solar and sidereal days, respectively. Let T_o represent the period of orbital revolution, that is, 1 year. As discussed above, the number of sidereal days in a year are one more than the number of solar days. Hence we obtain

$$\frac{T_o}{T_1} - \frac{T_o}{T} = 1 \, ,$$

which gives

$$\frac{1}{T_1} = \frac{1}{T} + \frac{1}{T_o} \, . \tag{1.2}$$

The period of revolution $P = 365.2563666$ mean solar days (T). Hence we obtain $T_1 = 0.99727$ solar days.

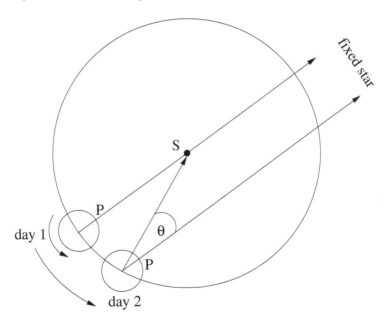

FIGURE 1.12: An observer P at Earth is directly below the Sun and a fixed star at noon on day 1. After completing 1 sidereal day, P returns to its position below the fixed star. The solar day is longer because P has to traverse an extra angle, θ, before the Sun is again directly overhead.

1.7   Astronomical Catalogs and Software

There exist a large number of catalogs, software, and web resources that provide information, such as positions, brightness magnitudes, proper motions, spectral classifications of stars, and other observable astronomical objects. The first star catalog was published by Ptolemy in the second century. It gave positions of 1,025 bright stars. A more modern catalog was published by the Smithsonian Astrophysical Observatory (SAO) in 1966. It lists a total of 258,997 stars brighter than magnitude 9. A more recent catalog is the Guide Star Catalog (GSC), which lists 945,592,683 stars up to magnitude 21 and is available online. Digitized sky maps are also available online from the STScI Digitized Sky Survey.

The task of identifying stars and constellations in the night sky is currently made very easy with the advent of many software programs that simulate the sky. One such software tool is Stellarium, available freely on the web. Using this, one can obtain a sky map at any time. This can be correlated with direct observations in order to identify the stars, planets, and constellations. I encourage the reader to use this software to become familiar with the night sky. You should try to identify constellations, bright stars, planets, etc. The Google Sky application is also very useful for the purpose of this identification.

Exercises

1.1 Use the Stellarium software to become familiar with the night sky. You should identify some of the prominent stars and planets visible in the night sky and try to visually locate them by direct observation.

1.2 Use Stellarium to learn about the shapes of different constellations in the night sky. By direct observation, try to identify a few of them. Try also to identify the Milky Way.

1.3 (a) The standard unit of angle is degrees and its subdivisions minutes and seconds. An alternate unit is radians, defined in Figure 1.13. Verify that $180° = \pi$ radians. Notice that for small angles, the length of the arc ABP is approximately equal to the chord AP. Hence θ is approximately equal to D/r for small angles.
(b) Another unit of angle is hours, the same as the unit of time. In this case, $360° = 24^h$ or $1^h = 15°$, where the superscript (h) denotes hours. Each hour is further subdivided into minutes and seconds in direct analogy with time units. For example, $50° = 3^h\ 20^m$, that is, 3 hours and 20 minutes. Convert $40°$ and $160°$ into time units (hours, minutes, and seconds).

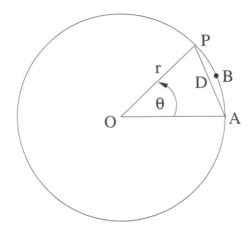

FIGURE 1.13: Let S denote the length of the arc ABP. The angle θ in radians is defined as $\theta = S/r$.

1.4 How was the mass and radius of the Earth measured historically? What is the best method to measure it at present?

1.5 How was the mass and radius of the Moon and the Sun measured historically? What is the best method to measure it at present?

1.6 Use Stellarium to observe the trajectories of different planets, the Sun, and the Moon. Notice that the trajectories of all the planets and the Moon lie roughly in a plane close to the ecliptic plane, the plane of apparent motion of the Sun.

1.7 Use Stellarium to see the retrograde motion of a planet such as Mars. You need to observe the motion of the planet with respect to fixed background stars. Hence you may observe the change in its position as you advance in units of a sidereal day. The retrograde motion is seen only when Mars approaches close to Earth. Hence you need to determine the time when this event happens. This can be easily done by experimenting with the software.

1.8 Verify that 1 sidereal day is equal to 0.99727 solar days.

1.9 In analogy with a light year, we may also define a light day and a light month as the distance light travels in a day and a month, respectively. Determine the values of these distances in terms of the Astronomical Unit (AU).

Chapter 2

Observations

We observe the Universe at many different wavelengths, visible light being most common. Observations are made mostly by ground-based telescopes. However, in order to eliminate atmospheric distortion, space-based telescopes, such as the Hubble Space Telescope, are also used. What we typically observe is the visible light emitted by different atoms and ions at the surface of stars. This gives us information about the temperature, relative density of different elements and their state of ionization at the stellar surface. We also observe the Universe at other wavelengths, such as radio, microwave, infrared, ultraviolet, x-rays, and gamma-rays. Excluding the radio, most of these observations are made using space-based telescopes, as these frequencies are strongly absorbed by the Earth's atmosphere. Besides this, we also observe cosmic rays, which are high-energy particles and consist mainly of protons and atomic nuclei. The highest energy cosmic ray particles are observed to have energies on the order of 10^{20} eV. This energy is extremely large. In comparison, the maximum energy of protons that has been achieved in the laboratory is on the order of 10^{13} eV. This was achieved in the Large Hadron Collider (LHC), which consists of a beam pipe of circumference 27 Km, equipped with the most powerful magnets operating at liquid helium temperatures. How does a cosmic site generate energies that are 10 million times higher? So far we do not have a satisfactory answer to this question.

2.1   Electromagnetic Waves

An electromagnetic wave consists of electric and magnetic fields propagating in space. The fields oscillate perpendicular to the direction of propagation. We can understand this with an analogy. Consider a wave propagating on a string in the z direction. The wave at three different times is shown in Figure 2.1. The string at any position moves up and down with time, that is, the displacement of the string is perpendicular to the z-axis, similar to the oscillation of the electric and magnetic fields in an electromagnetic wave.

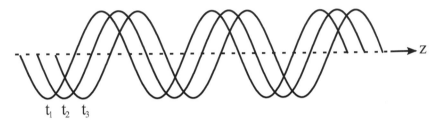

FIGURE 2.1: The displacement of a string at three different times t_1,t_2, and t_3.

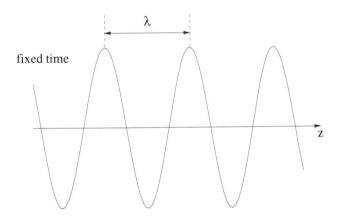

FIGURE 2.2: A snapshot of a traveling wave on a string at fixed time. The distance between two consecutive crests or displacement maxima is called the wavelength λ.

A snapshot of the string at any fixed time is shown in Figure 2.2. The distance between two maxima is called the wavelength, λ. The displacement of the string at any fixed position z at different times is shown in Figure 2.3. The time interval between two crests is called the time period T. In one time period, the particles on the string complete one full oscillation. The reciprocal or inverse of the time period is called the frequency, denoted by symbol ν. We can write this relationship as

$$\nu = \frac{1}{T}.$$ (2.1)

The frequency is interpreted as the number of complete oscillations per unit time. The unit of frequency is Hertz, denoted by Hz, which is equivalent to the number of oscillations per second. The speed of the wave is related to the frequency and wavelength by the formula

$$v = \nu\lambda.$$ (2.2)

The electromagnetic waves are more abstract in nature, representing the propagation of electric and magnetic fields, which cannot be directly visualized. We detect their presence indirectly through their influence on charged particles and magnets. These waves are also characterized by their frequency ν, which represents the number of oscillations of the electric and magnetic fields per unit time. The frequency is related to the time period in exactly the same manner as in the case of a wave on a string. Similarly, the wavelength represents the distance between two adjacent maxima of the electric field at any fixed time. The speed of an electromagnetic wave in vacuum is denoted by the symbol c, and is equal to

$$c = 299,792,458 \ \text{m/s}.$$ (2.3)

This is a very high speed, in fact the largest speed possible. No material particle or information can travel faster than the speed of light.

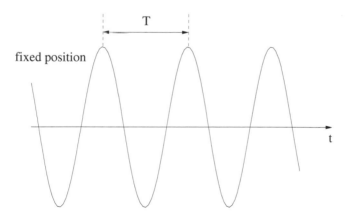

FIGURE 2.3: A traveling wave on a string shown as a function of time at a fixed position. The time interval between two successive crests or displacement maxima is called the time period.

In quantum mechanics, an electromagnetic wave is described as a beam of particles called photons. Each photon moves at speed c in vacuum. Its energy, E, is related to its momentum, p, by the formula

$$E = pc.   \tag{2.4}$$

Furthermore, its energy is also related to its frequency as

$$E = h\nu,   \tag{2.5}$$

where h is Planck's constant,

$$h = 6.6261 \times 10^{-34} \text{ Kg m}^2/\text{s} = 4.1357 \times 10^{-15} \text{ eV} \cdot \text{s}.   \tag{2.6}$$

We also define the constant $\hbar = h/2\pi$.

2.2   Electromagnetic Spectrum

Most of our information about the Universe is based on observations of electromagnetic radiation emitted by astronomical sources. We receive radiation over a wide range in the electromagnetic spectrum, from very low frequency radio waves to extremely high frequency gamma (γ) rays . The electromagnetic spectrum is classified into γ-rays, x-rays, ultraviolet, visible, infrared and radio, as shown in Figure 2.4. The γ-rays, x-rays, and ultraviolet (UV) radiation all have smaller wavelengths and higher frequencies in comparison to visible light. On the other hand, the infrared (IR), microwaves, and radio waves have larger wavelengths and hence smaller frequencies. The radio waves have the largest wavelengths and play a very important role in communications. In the lower panel in Figure 2.4, we also show the absorption of radiation by the Earth's atmosphere at different frequencies. We see that the gamma rays, x-rays, and UV radiation are completely absorbed by the atmosphere. As we reach visible wavelengths, the atmosphere becomes remarkably transparent. The absorption suddenly falls and becomes close to 0% in the entire visible range. The absorption due to atmosphere again starts increasing as we move to IR frequencies. It is interesting that the human eye has adapted itself to precisely that part of the spectrum that suffers minimal absorption due to atmosphere. At IR frequencies we see several frequency bands where the absorption is relatively small. Then as we reach the microwave frequencies, the absorption by atmosphere is again very large. Finally we see a wide window in radio waves where the atmosphere again becomes very transparent. This gets terminated at wavelengths larger than about 10 m due to the ionosphere.

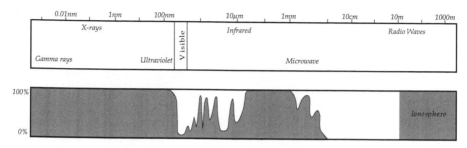

FIGURE 2.4: The electromagnetic spectrum from very small wavelength gamma rays (γ-rays) to high wavelength radio waves. The wavelength of different parts of the spectrum is also shown. The lower figure shows the absorption due to atmosphere at different wavelengths.

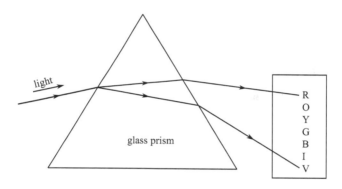

FIGURE 2.5: The electromagnetic spectrum of visible light as revealed by a prism. A beam of white light is incident on one face of the prism. The light emerging from the other face breaks up into different colors that are broadly classified into violet, indigo, blue, green, yellow, orange, and red.

The wavelengths of the visible spectrum lie roughly in the range 390 to 780 nm. The corresponding range of frequencies is 7.2×10^{14} to 3.8×10^{14} Hz. These are broadly classified into seven colors violet, indigo, blue, green, yellow, orange, and red. The violet and red colors have the smallest and largest wavelengths respectively. One can also see this spectrum visually using a prism, as shown in Figure 2.5. As the light rays enter the first face of the prism, they undergo refraction. This phenomenon is illustrated in Figure 2.6. This figure shows an electromagnetic wave propagating from medium 1 to medium 2, which may be air and glass, respectively. The interface is assumed to be a plane surface. The light rays bend as they propagate from one medium to another. The amount of bending is determined by the refractive index of the medium. For air, the refractive index is very close to 1, for glass it is about 1.5. Let us assume that light is incident at an angle θ_1 relative to the normal to the surface, as shown in Figure 2.6. The angle θ_2 at which it propagates in medium 2 is given by

$$n_2 \sin \theta_2 = n_1 \sin \theta_1 , \tag{2.7}$$

where n_1 and n_2 are the refractive indices in medium 1 and medium 2, respectively. The refractive index also depends on the frequency of light. If the incident light is a mixture or superposition of several different frequencies, these different components bend by different amounts. The refractive index in most materials increases with frequency. Hence a light beam of violet color undergoes larger refraction in comparison to red color.

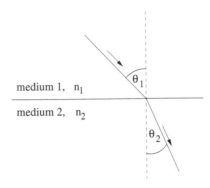

FIGURE 2.6: A beam of light is incident at an angle θ_1 with respect to the normal (dashed line) to the surface. As it propagates from medium 1 to 2, it bends, such that in medium 2 it makes an angle θ_2 to the normal. This is called refraction of light.

In the case of a prism, a light wave undergoes refraction at two surfaces, as shown in Figure 2.5. Let us assume that the incident wave is the standard white light, which is a mixture of all colors. At the left surface, the ray enters glass from air. Different colors undergo different amounts of refraction and hence the light beam breaks up into different colors. The light beam undergoes another refraction at the second surface, as it propagates from glass to air. The light emerging from this surface clearly shows seven different colors of the rainbow. This breaking up of light into different frequencies or colors is called dispersion.

We observe the radiation from astronomical sources using a wide range of instruments, both Earth and space based. The Earth-based observations are limited by the atmosphere. The absorption allows observations only in visible, radio and partially in IR. We see a remarkably low absorption in the visible region. In radio we also see a wide window from 300 MHz to 10 GHz that corresponds to a wavelength range of a few centimeters to about 10 m. At smaller frequencies, the ionosphere cuts off the radiation. In the infrared, we also see small windows where the atmosphere allows significant transmission.

2.3 Telescopes

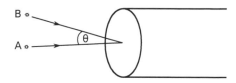

FIGURE 2.7: Two distant point objects at an angular separation θ viewed by a telescope. The image of each object gets smeared due to several effects, as shown in Figure 2.8.

FIGURE 2.8: The smeared image of two point objects shown in Figure 2.7. We can resolve these two objects only if their image is clearly separated.

A telescope has two basic components: an objective and an eye piece. The objective receives light from the astronomical object and forms its image. This is the most important component of a telescope. It may consist of a convex lens or a concave mirror. The two most important qualities that we seek in a telescope are its Light Gathering Power (LGP) and its resolving power. The LGP determines the brightness of the image. The resolving power is the ability of a telescope to distinguish two objects that may be very close to one another. For example, consider two objects, A and B, that may be located in the night sky at very small angular separation θ, as shown in Figure 2.7. For simplicity, let us assume that they are both point objects. As we view these objects through a telescope, their image gets slightly smeared (see Figure 2.8) for reasons explained later in this chapter. Hence point objects do not appear as point objects in the image. If the angle θ is very small, the image of these two objects will overlap. The minimum angle at which we can see them apart is called the resolution of the telescope. If the angular separation between two objects is less than this minimum angle, their image will overlap and we will not be able to resolve these objects. For larger angular separations, we will find these two objects clearly separated.

The image formed by a telescope can be viewed by an observer through the eye piece as shown in Figure 2.9. The image seen by the observer appears magnified. This property depends crucially on the choice of eye piece, as we explain below. There are two basic types of telescopes: (1) refractor telescope and (2) reflector telescope.

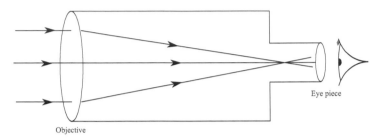

FIGURE 2.9: The two main components of a telescope are the objective and the eye piece.

2.3.1   Refractor Telescope

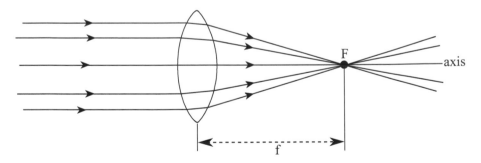

FIGURE 2.10: A parallel beam of light is focused at the focal point F by a convex lens.

A refractor telescope uses a convex lens as an objective. A convex lens, schematically shown in Figure 2.10, forms its image through refraction. It is usually made of glass. In Figure 2.10 we have shown its side view. The solid line passing through its center is called the axis. A parallel beam of light, incident on one of its surfaces, is focused at a point on the axis, denoted as F in Figure 2.10. This point is called the focal point and its distance from the center of the lens is called the focal length (f). The light rays undergo refraction twice as they pass through the lens: (1) as they enter the lens from air and (2) as they leave the lens, propagating from glass to air. In both cases, the rays are bent toward the axis. A plane passing through the focal point and parallel to the plane of the lens is called the focal plane. A parallel beam of light, which may be inclined with respect to the axis, is focused at a point on the focal plane, as shown in Figure 2.11. An astronomical object, such as a star, is located at very large distance from us and appears almost like a point. Due to the large distance, rays of light received from such an object are nearly parallel and hence its image is formed on the focal plane.

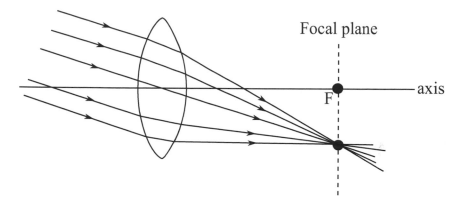

FIGURE 2.11: A parallel beam of light that is inclined at an angle with respect to the axis, is focused at a point on the focal plane.

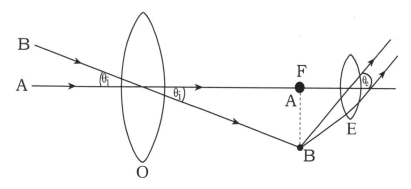

FIGURE 2.12: Here the left and the right lens are the objective (O) and eye piece (E), respectively. The objective forms the image of two astronomical objects A and B at its focal plane. This image, viewed through the eye piece, appears magnified and brighter.

The image formed by the objective can be viewed by eye through the eye piece (see Figure 2.12). The eye piece consists of another convex lens. It is positioned such that F is also located at its focal point. As explained above, a parallel beam of light is focused by a convex lens on its focal plane. Conversely, if an object is kept at the focal point (or focal plane) of a convex lens, then the rays from this object will form a parallel beam after emerging from such a lens. Hence, while viewing through the eye piece, the eye would receive a parallel beam of light exactly as it would receive if it were viewing the object without a telescope. The object would be perceived to be located at very large, nearly infinite distance. The advantage of viewing through a telescope is that the image is brighter and magnified.

Brightness: The fact that the image formed by a telescope is brighter is easily understood. It collects light over a larger area, the area of the objective, in comparison to the eye. The light is then focused over a small area, producing a bright image. This LGP is a very important property of a telescope. It is proportional to the area of the objective, or equivalently the square of its radius (R), that is,

$$LGP \propto R^2 . \tag{2.8}$$

Hence if we increase the radius of the objective by a factor of 2, the LGP increases by a factor of 4.

Magnification: The telescope also produces a magnified image of astronomical objects. Consider two point-like objects, such as stars. Assume that the angle that they subtend on a telescope is θ_i. Let the angle subtended on the eye, viewing through the eye piece, be θ_e (see Figure 2.12). The magnification produced by the telescope is defined as

$$m = \frac{\theta_e}{\theta_i} . \tag{2.9}$$

A large value of θ_e implies that the two objects subtend a larger angle at the eye and hence appear wider apart. From Figure 2.12 we see that

$$\frac{\tan \theta_e}{\tan \theta_i} = \frac{f_o}{f_e} ,$$

where f_o and f_e are the focal lengths of the objective and the eye piece, respectively. Typically we are interested in observing objects at very small angular separations θ_i. Hence both θ_i and θ_o are usually very small. For small angles we can use $\tan \theta \approx \theta$ and obtain

$$m \approx \frac{f_o}{f_e} . \tag{2.10}$$

Chromatic aberration: A major problem with refracting telescopes is that the image at different frequencies is formed at different positions. Hence the image gets distorted. This phenomenon is called chromatic aberration. The reason for this can be easily understood. The image is formed by refraction due to the objective lens. Hence, just as in the case of a prism, light rays emerging from a lens break up into different colors that form images at different positions. This aberration effect limits the quality of images that can be obtained with a refracting telescope.

2.3.2 Reflecting Telescope

A reflecting telescope uses a concave mirror as the objective, which forms its image due to reflection. A concave mirror is parabolic in shape. This has the property that a parallel beam of light is focused at a point on the axis of the mirror, called the focal point, as shown in Figure 2.13. A parallel beam of light that may be inclined to the axis is focused at a point on the focal plane, analogous to the image formed by a convex lens. The image of an astronomical object, formed by a concave mirror, is located on the same side as the object, in contrast to the image formed by a convex lens.

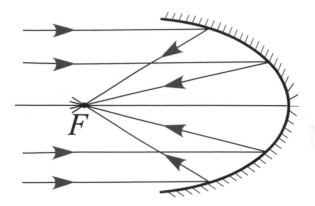

FIGURE 2.13: A parabolic concave mirror focuses a parallel beam of light at a point F on its axis called the focal point.

The image formed by the objective, also called the primary mirror, can be imaged or viewed by several different methods, as illustrated in Figures 2.14, 2.15 and 2.16. In the case of Prime Focus, Figure 2.14, the camera or a digital imaging device can be placed directly at the focal point of the primary mirror. This can be operated remotely. In all other cases, one uses a secondary mirror to direct the light out of the telescope. In Newtonian focus, shown in Figure 2.15, the light is directed out through a hole in the side of the telescope. One can view the resulting image through an eye piece, as illustrated in Figure 2.17. Finally, in Cassegrain focus, shown in Figure 2.16, the light is directed out through a hole in the primary mirror.

The reflecting telescopes offer several advantages in comparison to the refracting telescopes. One major advantage is that they do not suffer from chromatic aberration. Hence they are able to form much sharper images. Furthermore, it is possible to have much larger apertures in the case of reflecting telescopes. In the case of a refractor, a large aperture requires a very large convex lens. Even if one is able to manufacture a sufficiently large lens with the required precision, it is difficult to provide a support for it inside a telescope. This is partly because such large convex lenses are very bulky and also because the support would have to be inside the telescope. In contrast, the support for a reflecting concave mirror can be installed outside the telescope. Furthermore, a reflecting mirror can be easily segmented, thus reducing the requirements on the support system. Due to these advantages, most modern telescopes use a reflecting mirror as the objective.

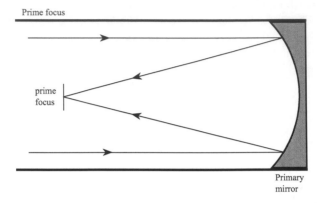

FIGURE 2.14: A reflector telescope used in the Prime focus mode. Here an imaging device, such as a camera, is placed directly at the focus of the primary mirror in order to record the image.

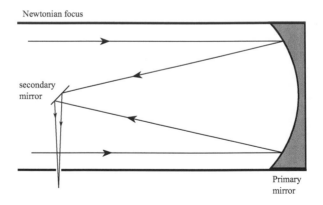

FIGURE 2.15: Newtonian focus: Here a secondary mirror is used in order to direct the light out of the telescope through an opening at the side of the telescope. The final image is formed outside the telescope.

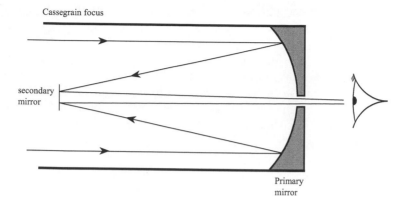

FIGURE 2.16: Cassegrain focus: Here a secondary mirror is used to focus the light out through an hole in the primary mirror.

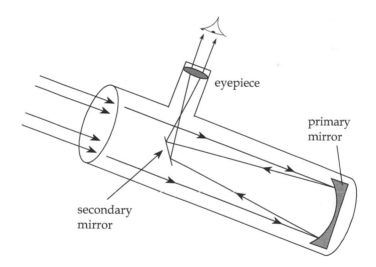

FIGURE 2.17: An example of a reflecter telescope using Newtonian focus. Here an eye piece is used to view the final image.

2.4   Observations at Visible Frequencies

At visible frequencies also, the atmosphere causes considerable distortion that limits the resolution. The intensity as well as the direction of starlight changes due to the atmosphere. The direction changes due to the refractive index of the atmosphere. Furthermore, light is scattered and absorbed by the atmosphere. Both of these effects cause attenuation of light and lead to reduced intensity at the Earth's surface. Scattering causes attenuation because it scatters the incident light beam in all directions. The light absorbed by the atmosphere is later emitted, but in directions different from the initial direction of propagation and at different frequencies. Hence the light received by an observer from a particular star is reduced due to both of these effects. The attenuation of light due to scattering and absorption is called extinction.

The atmosphere is always changing with time. The temperature, pressure, and wind velocity at any position show rapid fluctuations. Due to such fluctuations, the intensity and direction of starlight reaching Earth's surface also keep changing rapidly with time. The change in direction and intensity of light is called scintillation. This affects the stars more than the planets because stars appear as nearly point sources. The relevant quantity is the angular size of the object, which is the angle subtended by the star's surface at the position of the observer ($\Delta\theta \approx D/L$), where D is the diameter of the object and L its distance from Earth. Although stars have much larger diameters in comparison to planets, their distances from Earth are so large that their angular sizes are much smaller. Hence the fluctuations in their angular position due to scintillation are large in comparison to their angular size. Planets subtend a larger angle and hence the effect of scintillation is relatively small.

In order to understand the contribution of refraction, we assume a simple plane parallel model of the atmosphere, where different layers have different refractive indices, as shown in Figure 2.18. This model is valid as long as the star is located close to the vertical direction, as in this case we can neglect the curvature of Earth. Viewed from above the atmosphere, the star is located at an angle θ (see Figure 2.18). However due to atmospheric refraction it appears to be at an angle ξ. By applying Equation 2.7 to different layers, you may show as an exercise (Exercise 2.7) that

$$n_0 \sin\xi = \sin\theta,\qquad(2.11)$$

where n_0 is the refractive index of the layer of air close to the surface. The deviation of the ray, $\delta\theta = \theta - \xi << 1$, is given by

$$n_0 \sin\xi = \sin(\delta\theta + \xi),$$

and hence

$$\delta\theta \approx (n_0 - 1)\tan\xi \approx 58.2''\tan\xi,\qquad(2.12)$$

where the numerical result given in the last equation is obtained by observations and is valid for small values of ξ. For large ξ, this formula is not valid because the curvature of Earth is not ignorable. On the horizon we find that $\delta\theta \approx 35'$. Refraction leads to a shift in the observed direction of the star. If the atmosphere is stable, this observed direction will remain constant with time. However, due to fluctuations (convection, turbulence) the atmosphere is always changing. Therefore the observed direction keeps changing rapidly with time.

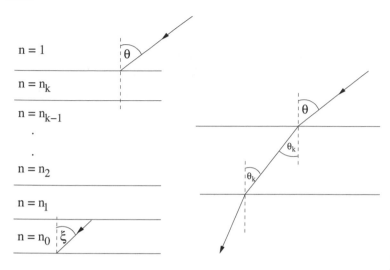

FIGURE 2.18: A plane parallel model of the atmosphere. We split the atmosphere into several layers with refractive indices n_0, n_1 ... n_k. Beyond the kth layer we assume that the refractive index is unity, corresponding to vacuum. A ray of light strikes the top of the atmosphere at an angle θ relative to the normal to the surface. After undergoing refraction through the atmosphere, it strikes the surface of Earth at an angle ξ. (Adapted from H. Karttunen et al., *Fundamental Astronomy*.)

2.4.1 Theoretical Limit on Resolution

A fundamental limit on the resolution of any instrument, which may be the human eye or a telescope, arises due to diffraction. The image of a point-like object formed by a telescope becomes smeared over a certain area due to this effect, as shown in Figure 2.19. Hence it is not possible to distinguish or resolve two objects if their smeared images overlap. This leads to a lower limit on the angular separation θ between two point objects that can be resolved by the telescope. The limit at wavelength λ is given by

$$\sin \theta \approx \theta \geq 1.22\lambda/D, \tag{2.13}$$

where D is the diameter of the aperture of the telescope. If the angular separation between two objects is smaller than this limit, then their images formed by the telescope will overlap. The resolving power increases with an increase in diameter; hence it is best to have as large a diameter or collection area as possible. Large aperture is also useful because it allows the instrument to capture a larger amount of the radiant energy emitted by the star. This is particularly important for detecting faint objects.

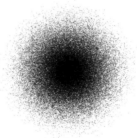

FIGURE 2.19: The image of a point-like object becomes smeared due to diffraction.

2.4.2   Seeing

One can, in principle, improve the resolution as much as possible by increasing the aperture diameter of a telescope. However, as the diameter becomes large, atmospheric effects lead to smearing of the image. This happens because the light received by different points on the aperture from a particular source travels through different paths in the atmosphere. Because the atmospheric fluctuations depend on position, these different rays undergo different amounts of refraction and hence form an image at different points. The image, therefore, is broken up into different spots or speckles, whose positions keep changing with time because the atmosphere is not stationary. Points sources, therefore, appear as vibrating speckles. This phenomenon is called seeing. This tends to obscure details in the source and we would like the resulting "seeing disk" to be as small as possible. Hence, although the theoretical resolution for a telescope with a large aperture may be small, in practice the resolution is limited by atmospheric fluctuations. This imposes a serious limitation on the resolution of Earth-based observations. The best resolution that can be achieved by Earth-based optical telescopes, that is, telescopes operating at visible wavelengths, is about $1''$. Some techniques such as speckle imaging can partially solve this problem. In this case, one takes a series of short exposure time images. The exposure time is taken to be sufficiently short so that over this time the atmosphere does not change significantly. The resulting images are then processed to produce a high-quality image. This technique has proved to be very successful for bright astronomical sources.

Some prominent ground based optical facilities are

(1) W. M. Keck Observatory: It is located near the summit of Mauna Kea in Hawaii at an altitude of 4,145 m. It has two telescopes, each of diameter 10 m, operating at optical and near-infrared frequencies.

(2) Very Large Telescope (VLT) (Paranal Observatory): It is located in the Atacama desert, Chile, at an altitude of 2,635 m. It consists of four separate telescopes, each of diameter 8.2 m, and operates at visible, near-, and mid-infrared frequencies.

2.5 Mounting of Telescope

The telescope must be mounted such that it has two rotation axes corresponding to the two degrees of freedom on the celestial sphere. While making an observation, one also needs to rotate it slowly and continuously in order to compensate for the Earth's rotation. There exist two standard methods for mounting the telescope: equatorial and azimuthal mounting.

2.5.1 Equatorial Mount

One of the axes (polar or hour axis) points along the axis of rotation of the Earth. The other axis (declination axis) lies on the equatorial plane. The telescope is continuously rotated about the polar axis to compensate for the Earth's rotation. This is discussed in more detail in the next chapter.

2.5.2 Azimuthal Mount

One of the axes points along the local vertical and the other along the horizontal. Here one needs to rotate about both axes in order to account for Earth's rotation.

The equatorial mount is convenient because one needs to rotate only about one axis in order to compensate for the Earth's rotation. However, the azimuthal mount is simpler and more stable because one of the axes points along the local vertical.

2.6 Interferometry

In this case we use two or more telescopes, or antennae in the case of radio waves, widely separated from one another. The waves received by different antennae are added together. This phenomenon of adding two waves is called interference. In Figure 2.20 we show a two-antenna interferometer. In radio astronomy, the waves received by the two antennae are first converted into electrical signals before adding. The best resolution achieved in this case is

$\theta \approx \lambda/D$, where D is the distance between the antennae and is called the baseline. Hence instead of using one huge antenna of diameter D, we can simply use many smaller antennae widely separated from one another. These sample the wave at different positions and produce a resolution that is determined by the distance between antennae. The technique has proven very useful for radio astronomy where the wavelength is much larger compared to optical wavelengths and hence we need a much larger value of D in order to achieve good resolution. Some large baseline radio telescopes are

(1) Very Large Array (VLA), USA, which consists of an array of 27 radio telescopes in a movable Y configuration. The diameter of each aperture or dish is 25 m, and the maximum configuration diameter is 35 Km. It is sensitive to wavelengths larger than 1.3 cm and has a resolution of $0.1''$.

(2) Very Large Baseline Interferometry (VLBI) uses radio telescopes scattered all over the Earth and hence can achieve very high resolution.

(3) Giant Metrewave Radio Telescope (GMRT), located in India, is the world's largest radio telescope at meter wavelengths. It has 30 antennae each with a dish diameter of 45 m, spread over an area of about 25 Km^2. The largest baseline is 28 Km.

The largest single dish radio telescope is located at Arecibo Observatory in Puerto Rico. It has a dish diameter of 305 m. It was constructed in a depression formed by a karst sinkhole.

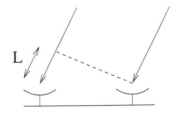

FIGURE 2.20: A two-antenna interferometer: The path difference between radiation from a distant source reaching the two antennae is L. The waves received by the two antennae are converted into electrical signals and added together.

2.7   Observations at Other Wavelengths

At wavelengths other than optical and radio, observations are best done above the atmosphere using satellites. Infrared Earth-based observations are possible in some frequency intervals. These are made at high altitudes in dry locations as water vapor absorbs infrared significantly. Some examples of Earth-based infrared observatories are (1) Mauna Kea Observatory at 4,205 m above sea level in Hawaii, (2) Infrared Observatory at Ladakh, India, at 4,500 m above sea level. Space-based infrared observations were first made by the Infrared Astronomical Satellite (IRAS) launched in 1983. One of its accomplishments was the detection of dust in orbit around young stars. This dust is expected to later form planets. Similarly, satellite observations have been made in UV, EUV, x-rays, gamma rays, etc.

Gamma rays are photons with energies $E_\gamma > 10^5$ eV. These are produced by atomic nuclei and hence probe the phenomenon of nuclear transitions in astrophysics. An interesting phenomenon in gamma radiation is the observation of γ-ray bursts. These were discovered by satellites sent by the United States to monitor nuclear tests in Soviet territory during the Cold War era. Instead, they found a large number of γ-ray bursts coming at random from different directions in the sky. The bursts lasted only for a few seconds. After the burst, the region of the sky is quiet. For a long time it was not even possible to identify any quiet source at the position of the burst. Recent observations have identified a faint optical source in some cases. This is identified as the galaxy in which the burst originated. Optical afterglows have also been observed in some cases. These objects are very far away from us. The energy released by a gamma ray burst in a few seconds is larger than the energy released by the Sun during its entire lifetime of approximately 10^{10} years. One possible explanation for the bursts is that they are caused by the collapse of a very massive, gigantic star, which leads to a very energetic supernova explosion.

Exercises

2.1 The visible spectrum lies in the wavelength range 390 to 700 nm. Determine the corresponding range of frequencies. Also determine the frequency of radio waves with wavelength 10 cm.

2.2 The Hubble Space Telescope has a primary mirror of diameter 2.4 m. Determine its angular resolution at a visible wavelength of 500 nm. Compare this with the best resolution of $1''$ possible with Earth-based optical telescopes.

2.3 Determine the resolution of the Very Large Array (VLA) at a wavelength of 2 cm, assuming a baseline of 35 Km. Determine also the resolution of a Very large baseline Interferometer (VLBI), that has a baseline of 1,000 Km at 2 cm wavelength.

2.4 Determine the energy in eV of a radio, visible, x-ray, and γ-ray photon. Assume that their wavelengths are 10 cm, 500 nm, 0.1 nm, and 10^{-5} nm, respectively.

2.5 Verify the conversion of Planck's constant, h, from Kg m^2/s to eV·s, given in Equation 2.6.

2.6 Using Equation 2.12, determine the deviation of a light ray that appears to arrive at an angle of 10° relative to the normal.

2.7 Prove Equation 2.11 for a plane parallel model of the atmosphere, illustrated in Figure 2.18. Hint: Apply Equation 2.7 sequentially to different layers, starting from the top. For example, $\sin\theta = n_k \sin\theta_k$, $n_k \sin\theta_k = n_{k-1} \sin\theta_{k-1}$, etc. Eliminate $\sin\theta_k$, $\sin\theta_{k-1}$ ..., θ_1.

Chapter 3

Astrometry

Astrometry is the branch of astronomy that deals with the positions and motions of heavenly objects. The positions are specified in terms of their distances and angular positions on the sky. We need to choose a suitable reference or coordinate system for this purpose. For example, the entire surface of the Earth is mapped by longitudes and latitudes. The latitude and longitude of any city give us its angular position or coordinates. We similarly imagine a grand celestial sphere with all the stars and planets scattered on its surface. We are interested in defining a suitable angular grid to specify angular positions on this sphere. Besides the two angular coordinates, we also need the third coordinate, namely the distance of the object from some chosen point, called the origin of the coordinate system. The origin may be taken to be the position of the Earth or the Sun.

The positions, as observed from Earth, keep changing due to several factors. Dominant among these is the rotation of Earth around its axis. This causes a periodic change in the observed angular positions of the stars, which appear to move from east to west in the sky. It is useful to take this periodic shift out of our observations because we want to focus on the properties of the stars rather than the rotation of Earth. For this purpose we define a coordinate system that does not rotate along with Earth. Once this is done, the positions of all stars remain approximately fixed. You can verify this directly by observing the night sky or by using software such as Stellarium. As you observe the stars over a period of a few hours, you will see them move from east to west. However, their relative positions with respect to one another remain fixed. In contrast, the planets and the Moon show significant motion with respect to the background stars.

The positions of stars also change very slightly due to the revolution of Earth around the Sun. This shift arises due to a phenomenon called parallax, which we discuss in detail later in this chapter. Besides that, the positions of stars change slightly due to their intrinsic motion relative to the Sun. This motion as well as the shift due to parallax are so small that they can be detected only with sophisticated measurements. The planets, of course, move much more rapidly in elliptical paths around the Sun.

The position and velocity of any object are both vector quantities. A vector quantity has magnitude as well as direction. In contrast, quantities such as height, thickness, and radius are scalars and only have magnitude, but no direction. For example, if we wish to specify the position of a star, we need to give both its distance from us as well as the direction of its location. Hence it is a vector quantity and we call it the position vector of a star. Similarly, the velocity vector gives us information about the total speed of an object as well as its direction of motion.

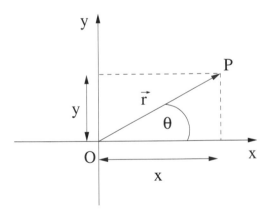

FIGURE 3.1: The position vector $\vec{r}$ of point P on a two-dimensional surface.

We next briefly discuss the concept of a vector. The position vector, $\vec{r}$, of a point P on a two-dimensional plane surface, labeled by coordinates (x, y), is shown in Figure 3.1. These are called the Cartesian coordinates of P. The vector, $\vec{r}$, makes an angle θ with the x-axis. Let the length of the vector, that is, the distance of P from O, be denoted by r. Its projections on the x- and y- axes are given by

$$x = r \cos \theta \,,$$
$$y = r \sin \theta \,. \tag{3.1}$$

These are called the components of the vector $\vec{r}$. An alternate set of coordinates is given by (r, θ). These are called the polar coordinates. For a general vector $\vec{V}$, we denote its magnitude by V and its components by V_x and V_y. If $\vec{V}$ makes and angle θ with the x-axis, we obtain

$$V_x = V \cos \theta \,,$$
$$V_y = V \sin \theta \,. \tag{3.2}$$

Here we are primarily interested in the velocity vectors of stars, besides their position vectors.

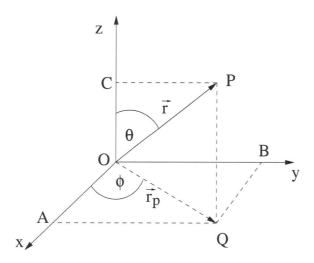

FIGURE 3.2: The position vector $\vec{r}$ of point P in a three-dimensional space.

In three dimensions, the position of any object is specified by three coordinates. We can choose these to be the Cartesian coordinates (x, y, z). The position vector $\vec{r}$ of point P is shown in Figure 3.2. In this figure, PC and PQ are perpendicular to the z-axis and the $x - y$ plane, respectively. Also, QA and QB are perpendicular to the x- and y- axes, respectively. The z component of $\vec{r}$ is equal to $z = r \cos \theta$. The magnitude of the projection vector, $\vec{r}_p$, is $r_p = r \sin \theta$. Using Equation 3.1, the x and y components of the vector $\vec{r}$ are $x = r_p \cos \phi = r \sin \theta \cos \phi$ and $y = r_p \sin \phi = r \sin \theta \sin \phi$. Hence we obtain

$$
\begin{aligned}
x &= r \sin(\theta) \cos(\phi) \,, \\
y &= r \sin(\theta) \sin(\phi) \,, \\
z &= r \cos(\theta) \,.
\end{aligned}
\tag{3.3}
$$

We can also use the coordinates (r, θ, ϕ) instead of (x, y, z). These are called the spherical polar coordinates. Here r, θ and ϕ are called the radial, polar, and azimuthal coordinates, respectively. The Cartesian coordinates, (x, y, z), can vary over all possible values from $-\infty$ to $+\infty$. The spherical polar coordinates, however, take values in the range $r \geq 0$, $0 \leq \theta \leq \pi$, $0 \leq \phi < 2\pi$, where all the angles are given in radians. In general, the x, y and z components of any vector $\vec{v}$ in three dimensions are given by

$$\begin{aligned}
v_x &= v\sin(\theta)\cos(\phi)\,, \\
v_y &= v\sin(\theta)\sin(\phi)\,, \\
v_z &= v\cos(\theta)\,,
\end{aligned} \qquad (3.4)$$

where v is the magnitude of the vector.

In astronomy, we essentially use the spherical polar coordinates. The only difference is that the latitude, δ, defined as

$$\delta = 90° - \theta\,, \qquad (3.5)$$

is used instead of the polar coordinate θ (see Figure 3.3). This coordinate takes values in the range $-90°$ to $90°$. The azimuthal coordinate ϕ is equivalent to the longitude.

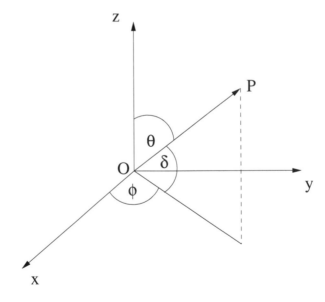

FIGURE 3.3: The angular coordinates (δ, ϕ) of a point P.

In many cases we are not interested in the radial coordinate or the distance of the object. The distance of astronomical objects is also much more difficult to measure in comparison with their angular coordinates. We can focus on the angular coordinates by projecting each object on an imaginary sphere with Earth as its center and very large (almost infinite) radius. This is called the celestial sphere.

The velocity vector $\vec{v}$ of an object P is schematically shown in Figure 3.4. Its magnitude, v, is called the speed. Its Cartesian components (v_x, v_y, v_z) are given by Equation 3.4. In astronomy, it is convenient to use the radial component, v_r, and the tangential component, $\vec{v}_t$, as shown in Figure 3.4. The radial component, v_r, is simply the component of $\vec{v}$ along the radial direction. The tangential component, $\vec{v}_t$, is the component in the tangential plane or the plane perpendicular to the position vector, $\vec{r}$, of P. Hence we can express the velocity vector as

$$\vec{v} = (v_r, \vec{v}_t). \qquad (3.6)$$

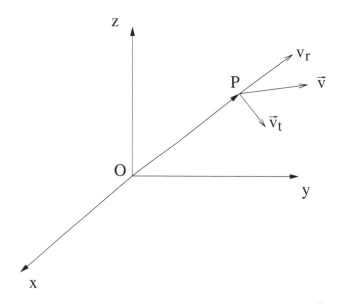

FIGURE 3.4: The radial, v_r, and tangential, $\vec{v}_t$, velocity components of the velocity vector, $\vec{v}$, of the object P.

3.1 Coordinate Systems

In order to specify the two angular coordinates we need to pick an origin and choose some reference for specifying the latitude and longitude of an object. The latitude is specified by choosing a reference plane or the equatorial plane, passing through this origin. This plane lies at latitude $\delta = 0$. The longitude or the azimuthal angle is specified by choosing some reference direction. Some of the standard coordinate systems are described below.

3.1.1   The Horizontal System

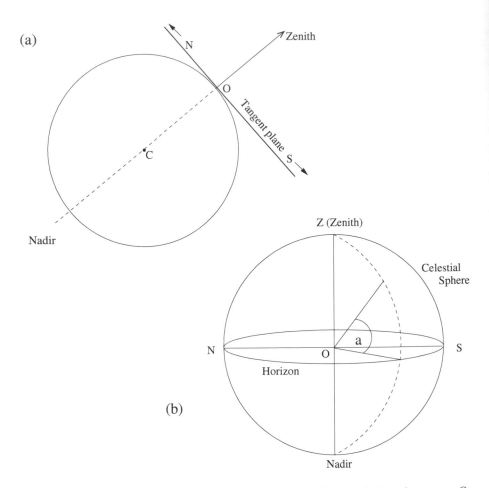

FIGURE 3.5: (a) The tangent plane at the position of the observer, O, located at the surface of the Earth. Here C denotes the center of the Earth. Also shown are the zenith and the nadir. (b) The celestial sphere with the observer at its center.

This coordinate system is centered at the observer, O, who may be located at some point on the surface of the Earth. We consider the tangent plane at the location of the observer, as shown in Figure 3.5a. This plane cuts the celestial sphere at the horizon. We shall refer to this plane as the horizontal plane. The points on the horizon toward north and south are denoted as N and S, respectively in Figure 3.5. The great circle NZS (Figure 3.5b) is called the meridian. All great circles passing through the zenith are called verticals. The position of any object in this coordinate system is specified by the altitude or elevation a and the azimuthal coordinate or azimuth A. Here the horizontal plane acts as the reference plane and the verticals as the longitudes. The altitude a is the angle along the vertical with respect to the horizontal plane. It lies in the range $[-90°, 90°]$ and is analogous to the latitude, Equation 3.5. The azimuth A is the angular position of the vertical passing through the object with respect to some fixed direction. There is no universal definition of this fixed direction. An observer chooses some convenient reference and A can be measured clockwise or counterclockwise.

This coordinate system is most convenient from the point of view of the observer. However, stars appear to move from east to west due to the rotation of the Earth. Hence the positions of stars change rapidly in this system, and it is not convenient for use in catalogs.

3.1.2 Equatorial Coordinate System

We need a system in which the positions of stars remain fixed with time, at least approximately. One such system is the equatorial coordinate system. Here we use the fact that the axis of rotation of Earth remains approximately fixed. It points about $1°$ away from the star Polaris ($1° \approx 2$ full moons). The equatorial plane that is perpendicular to this axis also remains fixed. We extend this plane so that it cuts the celestial sphere. The intersection is called the equator of the celestial sphere. This is taken to be the reference plane to specify the latitude of the object. This coordinate is called the declination δ (Dec). It lies in the range $[-90°, 90°]$, with points on the equatorial plane having $\delta = 0$. The second or azimuthal coordinate is called right ascension α (or RA). It is measured counterclockwise from a fixed point in the sky called the vernal equinox. Recall that in spring, the Sun appears to move from the southern to northern hemisphere. The time of this event, that is, when the Sun crosses the equatorial plane, as well as the direction to the Sun at that moment is called the vernal equinox (see Figure 1.5). The coordinate RA takes values in the range of 0 to $360°$, analogous to the longitude. It is also specified in time units, 0 to 24 hours. For example, $\alpha = 90°$ is equal to 6 hours, denoted as 6^h, in time units.

The stars remain approximately fixed in this coordinate system. They are unaffected by the rotation of Earth. If we choose the origin as the center of the Earth, then the angular positions of stars will change slightly with time due to Earth's revolution. This shift can be eliminated by choosing the center of the Sun as the origin. In any case, such small changes are irrelevant for most purposes. The axis of rotation of Earth also undergoes a slow change due to the precession, as discussed later in this chapter. Hence the equatorial frame changes slowly and one has to specify the time or epoch in which the system is used. The system is updated every 50 years. Before 2000, the system B1950 was in use. Now the system J2000 is being used.

3.1.3   Ecliptic System

Ecliptic is the orbital plane of the motion of the Earth around the Sun or, equivalently, it is the plane in which the Sun appears to move around the Earth (see Figure 1.4). The ecliptic pole is the point where a line perpendicular to the ecliptic at the origin meets the celestial sphere. The origin may be chosen to be the position of the Earth or the Sun. The equatorial and ecliptic planes intersect along a straight line directed toward the vernal equinox, which is the reference direction for the definition of RA, the azimuthal coordinate in the equatorial system. In the ecliptic system, the ecliptic plane is used as the reference for latitude. The reference for longitude is taken to be the vernal equinox in this case also.

3.1.4   Galactic Coordinate System

The galactic coordinate system is defined by using the fact that the Milky Way forms a planar structure. This plane is used as a reference plane to specify the galactic latitude b. The galactic longitude l is measured counterclockwise from the direction of the center of the galaxy. In equatorial coordinates, J2000, the galactic pole is located at $\delta = 27.13°$ and $\alpha = 192.86°$ and the galactic center is located at $\delta = -28.94°$ and $\alpha = 266.40°$.

3.1.5   Supergalactic Coordinate System

In this case the supercluster plane is used as the reference for defining the supergalactic latitude (SGB). The supergalactic pole is located at $\alpha = 283.8°$ and $\delta = 15.7°$ in equatorial J2000 system. The origin for the longitude (SGL) is taken to lie along the intersection of the galactic and supercluster plane. The reference for longitude lies approximately in the direction $\alpha = 42.3°$ and $\delta = 59.5°$, where α and δ are the equatorial coordinates.

3.2   Space Velocity and Proper Motion of Stars

The stars remain approximately fixed in the sky. Their relative positions, with respect to one another, change only slightly with time. This is attributed to their intrinsic motion. The shift is so small that it is detectable only over a long time interval with sophisticated instruments.

The velocity of a star with respect to the Sun is called the space velocity. We can decompose it into the radial and the tangential components. The radial component, v_r, is the velocity of the object along the line of sight, as shown in Figure 3.6. This is measured by the Doppler shift, explained below, and is relatively easy to measure. The tangential or transverse component,

$\vec{v}_t$, is perpendicular to the radial direction or the line of sight. This is the component tangential to the celestial sphere and can point in any direction on the tangent plane. Hence it is a vector with two components and is denoted by a vector symbol $\vec{v}_t$. We can, therefore, write the space velocity of a star as

$$\vec{v} = (v_r, \vec{v}_t).$$

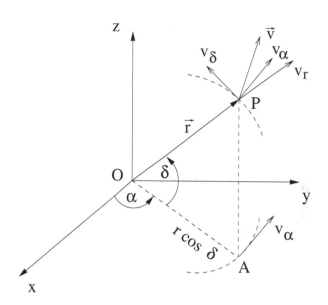

FIGURE 3.6: The position vector of object P is $\vec{r}$. It is located at coordinates (r, δ, α). Its space velocity is indicated by $\vec{v}$. The radial component, v_r, is the component in the direction of $\vec{r}$ and the tangential component, $\vec{v}_t$, lies in the plane perpendicular to $\vec{r}$. The vector $\vec{v}_t$ is resolved into two components, (v_δ, v_α). Drop a perpendicular PA on the $x-y$ plane. The component v_δ lies in the plane OPA and points in the direction of increasing δ at P. The component v_α is parallel to the $x-y$ plane and points in the direction of increasing α at P. For convenience, we also show this component projected on the $x-y$ plane at A. Hence the component v_δ is tangential to the longitude passing through P and points in the direction of increasing latitude. Similarly, v_α is tangential to the latitude passing through P and points in the direction of increasing longitude.

In astronomy, the entire sky is mapped in a grid of latitudes and longitudes. At any point we can decompose $\vec{v}_t$ in terms of its components along the local latitude and longitude, as explained in Figure 3.6. In equatorial coordinates, these two components are denoted as v_δ and v_α, respectively. Hence we can write, $\vec{v}_t = (v_\delta, v_\alpha)$. We are often interested only in the angular velocity of the star. This is the rate at which the angular position of a star changes with time and is called the proper velocity, denoted by $\vec{\mu}$. Let us assume that in a small time interval t, a star traverses a small angle Δ in the sky. During this time, it moves along a distance D across the sky. Clearly, $D = r\Delta$, where r is the distance of the star from the Sun and Δ is in radians. The angular speed of the star is $\mu = \Delta/t$ and the tangential speed $v_t = D/t$. Hence we find that $v_t = \mu r$. The proper velocity is also related to the tangential velocity by a similar relationship, $\vec{v}_t = \vec{\mu} r$. Let μ_δ and μ_α represent the rate of change of the angles δ and α, respectively, with respect to time, that is, $\mu_\delta = \dot{\delta}$ and $\mu_\alpha = \dot{\alpha}$. You can show as an exercise that the velocity components can be written as $v_\delta = r\,\mu_\delta$, $v_\alpha = r\cos\delta\,\mu_\alpha$. Hence we can write the proper velocity $\vec{\mu}$ as

$$\vec{\mu} = (\mu_\delta, \cos\delta\,\mu_\alpha)\,, \tag{3.7}$$

and the magnitude

$$\mu = \sqrt{\mu_\delta^2 + \cos\delta^2\mu_\alpha^2}\,. \tag{3.8}$$

The largest known proper motion is that of Barnard's star, which moves across the sky at a speed of $10.3''$ per year. This star turns out to be too faint to be observable by the naked eye. Its angular velocity is equivalent to traversing the angle subtended by the diameter of the moon in less than 200 years.

3.2.1 Doppler Effect

FIGURE 3.7: A source S moving away from an observer O at speed v emits light of wavelength λ_0. Due to the Doppler effect, the wavelength λ observed by O is larger than λ_0.

Consider a source, S, which is moving away from an observer O at speed v as shown in Figure 3.7. The source emits radiation at wavelength λ_0. The wavelength seen by the observer, $\lambda = \lambda_0 + \Delta\lambda$, turns out to be larger by an amount $\Delta\lambda \approx \lambda_0(v/c)$, where c is the speed of light. This relationship is valid for $v \ll c$. Hence the wavelength becomes larger (is red shifted) if the source is moving away and smaller (is blue shifted) if it is moving toward the observer. This shift in wavelength is called the Doppler effect. In general, the source may be moving in any direction, not necessarily along the line of sight. Let its velocity be denoted by $\vec{v}$. In this case the Doppler shift is determined by the component of $\vec{v}$ along the line of sight. Assuming that O is at the origin of the coordinate system, this is just the radial component v_r. Hence we obtain

$$\frac{\Delta\lambda}{\lambda_0} \approx \frac{v_r}{c}\,, \tag{3.9}$$

valid for $v_r \ll c$. This effect can be used to deduce the radial velocity v_r of a star. We discuss this in Chapter 7. In general, it is much easier to measure the radial velocity of a star in comparison to the angular velocity.

3.3 Parallax

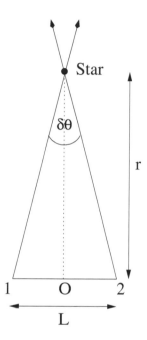

FIGURE 3.8: The shift $\delta\theta$ in the angular position of a star as the observer O moves from position 1 to 2. Here L denotes the distance moved by the observer perpendicular to the line of sight and r is the distance to the star.

If we observe an object from different positions, as illustrated in Figure 3.8, we see it in different directions. The difference in the observed direction or the shift in the angular position of the object is called the parallax. All stars and galaxies are located at very large distances from Earth in comparison to typical distances within the solar system. For example, the distance of the nearest star, beyond the Sun, is about 200,000 times larger in comparison to the Earth-Sun distance. Due to this large distance, the parallax observable for a star is very small. The largest parallax arises due to the revolution of Earth. Even this is very small and can be detected only by very precise measurements. Furthermore, this measurement is possible only for stars close to us. The shift is undetectable for stars that are very far away.

Let r be the distance of a star and L the distance moved by the observer perpendicular to the line joining the star to the observer, as shown in Figure 3.8. Then the parallax is the angular shift $\delta\theta$,

$$\delta\theta = \frac{L}{r}. \tag{3.10}$$

This relationship is valid for small angles. The distance L is also called the baseline. By measuring $\delta\theta$, we can estimate the distance of the star.

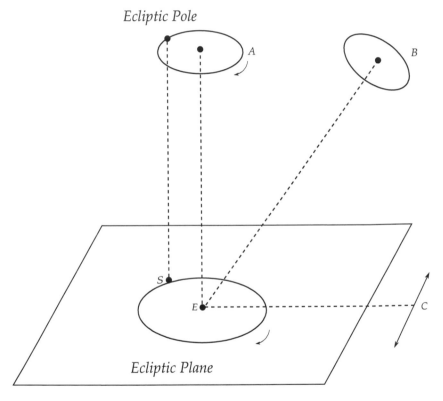

FIGURE 3.9: The Sun (S), as observed from Earth (E), appears to move in a circle in the ecliptic plane. A star A which lies directly above the Sun along the line perpendicular to the ecliptic plane also appears to move in a circle. A star B at any other position appears to move in an ellipse. A star C which lies in the ecliptic plane, shows a straight line motion.

In order to measure astronomical distances, we need large baselines, such as the Earth's orbit around the Sun. What is measured is the shift in the angular position of a star with respect to the distant background stars due to the annual motion of the Earth. This is convenient because the angular positions of the distant stars remain approximately fixed. Hence the shift in the relative angular position of a star with respect to a distant star directly gives a measure of parallax. This method is useful for stars at distances of less than about 100 pc.

Due to the annual motion of Earth, a star appears to move in an elliptical orbit. The eccentricity of the ellipse is determined by the angular position of the star, as shown in Figure 3.9. The angle subtended by the semi-major axis of this ellipse is called the annual parallax π (Figure 3.10). It is the same as the angle subtended by the orbital radius of Earth (1 AU) at the star. If $\pi = 1''$, then the distance to the star is defined to be 1 *parsec* (pc). Because 1 *radian* = $206,265''$, this implies that if the angle subtended $\delta\theta = 1''$, then the distance

$$r = \frac{1 \text{ AU}}{\delta\theta} = 206,265 \text{ AU}.$$

Hence we find that 1 pc = 206265 AU = 3.26 light years.

The stars themselves are in motion with respect to the Sun. Therefore, the observed parallax of a star also depends on its proper motion μ. The distance as well as μ are both extracted simultaneously by parallax measurements. The proper motion of all stars is observed to be approximately independent of time. Hence this gives a contribution to parallax that increases linearly with time (t) in a fixed direction, that is, $\delta\theta = \mu t$. In contrast, as discussed above, the star appears to move in an ellipse due to the annual motion of the Earth. Hence this contribution does not show a linear dependence on time and also changes direction. In particular, it becomes zero after one complete year. Due to the different time dependences of these two components, they can be separated by making a large number of measurements at different times. Hence we can extract both the proper motion and the distance of a star.

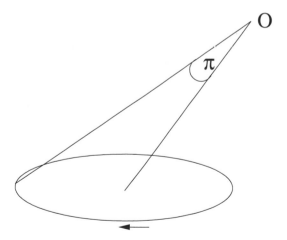

FIGURE 3.10: Due to the annual motion of the Earth, the stars appear to move in an ellipse. The parallax π is the angle subtended by the semi-major axis of this ellipse at the position of the observer O.

The parallax method is the only direct method for measuring astronomical distances. All other methods rely on some physical property of a star, through which we can deduce its luminosity. Hence the extracted distance may depend on assumptions about its physical properties. At visible frequencies, using Earth-based telescopes, this technique has been used to measure distances on the order of 50 to 100 pc. For larger distances, the angular shifts become too small to be measurable due to distortions caused by the atmosphere and other effects. Recent measurements at microwave frequencies using Very Long Baseline Interferometry (VLBI) have extended this range to distances on the order of 1 Kpc. A futuristic space-based observatory, Space Interferometry Mission (SIM), plans to extend the distance measurement at visible wavelengths to 1 Kpc using direct parallax.

3.4   Aberration

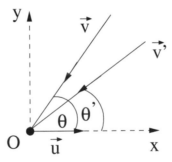

FIGURE 3.11: The angular position of an object appears shifted to an observer in motion due to the aberration effect. Here the observer O is moving at velocity $\vec{u}$ along the x-axis. In the rest frame the star is located at an angle θ and the velocity vector of light emitted by the star is $\vec{v}$. However due to the motion of O, she will observe the velocity vector of light to be $\vec{v}' = \vec{v} - \vec{u}$. This leads to a shift in the angular position of the star.

Another effect that leads to a shift in the angular positions of stars is aberration. Anyone can experience this phenomenon by taking a brisk walk or by riding a bicycle in rain. Let us assume that an observer at rest sees the rain drops falling vertically downwards. They would appear to strike at an angle to an observer in motion. She will need to tilt her umbrella slightly forward in order to avoid getting wet. The situation can be understood by considering relative velocity. Consider an observer O moving at velocity $\vec{u}$ along the x-axis, as shown in Figure 3.11. Let the true velocity of rain be $\vec{v}$. This is the velocity observed by O if she was at rest. However since O is in motion, she will observe the velocity to be $\vec{v}' = \vec{v} - \vec{u}$. As shown in Figure 3.11, $\vec{v} = -v\cos\theta\hat{x} - v\sin\theta\hat{y}$, and $\vec{u} = u\hat{x}$. Hence the angle θ' that $\vec{v}'$ makes with the x-axis is given by

$$\tan\theta' = \frac{\sin\theta}{\cos\theta + \frac{u}{v}} \qquad (3.11)$$

In application to stars, we simply need to replace the rain with starlight. Hence if the position of the star in the rest frame is at an angle θ with respect to the direction of motion of the observer, the observed position θ' is given by Equation 3.11 with v replaced by c, the velocity of light. The aberration effect also causes small angular shifts in the positions of stars due to the annual motion of the Earth. It is clear that the effect arises due to finite speed of light and the shift is of order u/c for small u.

3.5   Coordinate Transformations

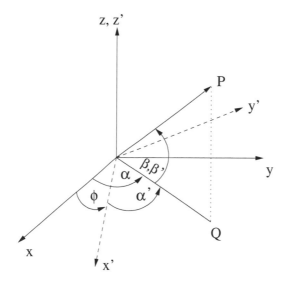

FIGURE 3.12: The coordinate system, (x', y', z'), is obtained by a counter-clockwise rotation of the system, (x, y, z), about the z-axis by an angle ϕ. The angular coordinates of the point P in the (x, y, z) and (x', y', z') systems are (β, α) and (β', α'), respectively.

We next consider the transformation among different coordinate systems. Consider an object P, whose angular coordinates are (β, α) and (β', α') in the coordinate systems (x, y, z) and (x', y', z'), respectively. Here β, β' denote the latitudes and α, α' the azimuthal angles. We are interested in determining the relationship between these angular coordinates. In general, the system (x', y', z') is related to (x, y, z) by three angles, which can be chosen to be the Euler angles. This is discussed in more detail below. Here we first consider some special cases.

Let's first assume that the system (x', y', z') is related to (x, y, z) by a counterclockwise rotation ϕ about the z-axis, as shown in Figure 3.12. In this case it is clear that

$$\beta' = \beta, \qquad \alpha' = \alpha - \phi. \tag{3.12}$$

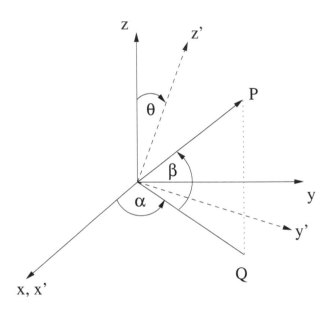

FIGURE 3.13: The coordinate system, (x', y', z'), is obtained by a clockwise rotation of the system (x, y, z) about the x-axis by an angle θ.

Next we consider a clockwise rotation by an angle θ about the x-axis as shown in Figure 3.13. In this case the coordinates of any point P in the two systems are related by

$$
\begin{aligned}
x' &= x, \\
y' &= \cos\theta\, y - \sin\theta\, z, \\
z' &= \sin\theta\, y + \cos\theta\, z.
\end{aligned} \tag{3.13}
$$

In terms of the angular coordinates, we have,

$$
\begin{aligned}
x &= \cos\beta \cos\alpha, \\
y &= \cos\beta \sin\alpha, \\
z &= \sin\beta,
\end{aligned} \tag{3.14}
$$

with the corresponding relationships for the primed coordinates. Here we are only interested in the angular coordinates and hence have set the distance of the source equal to unity. Hence we obtain

$$\begin{aligned}
\cos\beta'\cos\alpha' &= \cos\beta\cos\alpha, \\
\cos\beta'\sin\alpha' &= \cos\theta\cos\beta\sin\alpha - \sin\theta\sin\beta, \\
\sin\beta' &= \sin\theta\cos\beta\sin\alpha + \cos\theta\sin\beta.
\end{aligned} \tag{3.15}$$

We can use these relationships to extract β' and α'. The angle β' can be extracted directly by using the third equation in Equation 3.15. We can next use the first two equations to extract α'. Both of these equations are required to unambiguously fix the quadrant in which this angle lies.

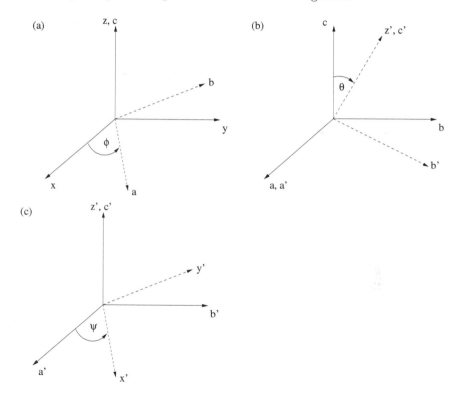

FIGURE 3.14: The three Euler rotations for transformation from coordinate system (x, y, z) to (x', y', z').

In the most general case, the system (x', y', z') is related to (x, y, z) by the three Euler angles, as shown in Figure 3.14. We first obtain the system (a, b, c) by a rotation about the z-axis by an angle ϕ. Next we make a rotation about the a-axis by an angle θ to obtain the system (a', b', c') and finally a rotation about the c'-axis by an angle ψ to obtain (x', y', z'). The angles ϕ, θ, ψ are called the Euler angles. We notice that each of these three transformations correspond to the two special cases discussed above. Hence we can obtain the general transformation using the results of these special cases.

Let's now use the Euler angles to determine the transformation between the equatorial, (x, y, z), and the galactic, (x', y', z'), coordinates. We will only indicate the steps required without giving detailed formulae. In the equatorial coordinates, the galactic pole (z') and the galactic center (x') are located at $(\delta = 27.13°, \alpha = 192.86°)$ and $(\delta = -28.94°, \alpha = 266.40°)$, respectively. We first determine the line of intersection of the (x, y) and (x', y') planes. This line is perpendicular to both the z- and z'-axes. Let us assume that this line is located at an angle ϕ from the x-axis (in $x - y$ plane). We rotate about the z-axis to align the x-axis with the line of intersection, giving us the new coordinate system (a, b, c) as shown in Figure 3.14(a). Here the c-axis, the polar axis of the new frame, coincides with the z-axis. We determine the location of the galactic pole and the galactic center in system (a, b, c). Let the angle between z- or c-axis and the z'-axis be θ. The a-axis is perpendicular to both the c-axis and z'-axis. Hence a rotation about the a-axis by θ aligns the c-axis with the z'-axis, as shown in Figure 3.14(b). The new (a', b') plane is now also aligned with the (x', y') plane. Determine the location of galactic center (x') in the (a', b', c') coordinates. Let it make an angle ψ with the a'-axis. Then a rotation about the c'-axis by angle ψ gives us the desired transformation, as shown in Figure 3.14(c).

3.5.1 Transformation between Equatorial and Ecliptic Coordinate Systems

We next explicitly determine the transformation between the equatorial and the ecliptic coordinate systems. These two coordinate systems use the same reference, the vernal equinox, for the azimuthal angle. The ecliptic pole is aligned at an angle $\theta = 23°26'$ with respect to the equatorial pole, as shown in Figure 3.15. We are interested in finding the ecliptic angular coordinates (β, λ), given the equatorial coordinates (δ, α). Here β, λ denote the ecliptic latitude and longitude, respectively. The relationship is given by Equation 3.15 with the following substitutions

$$\beta' \to \beta, \quad \alpha' \to \lambda, \quad \beta \to \delta, \quad \alpha \to \alpha, \quad \theta \to -\theta.$$

We need to change $\theta \to -\theta$ because we need to rotate counterclockwise about the vernal equinox, as shown in Figure 3.15. These substitutions give us

$$
\begin{aligned}
\cos \beta \cos \lambda &= \cos \delta \cos \alpha, \\
\cos \beta \sin \lambda &= \cos \theta \cos \delta \sin \alpha + \sin \theta \sin \delta, \\
\sin \beta &= -\sin \theta \cos \delta \sin \alpha + \cos \theta \sin \delta,
\end{aligned}
\tag{3.16}
$$

from which we can determine β and λ.

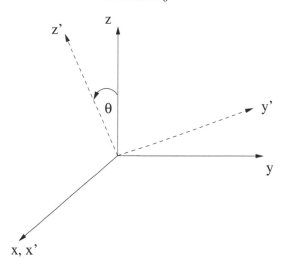

FIGURE 3.15: Transformation between the equatorial (x, y, z) and the ecliptic (x', y', z') coordinate systems. The x- and x'-axes both point toward the vernal equinox. The transformation is obtained by a counterclockwise rotation by angle $\theta = 23°26'$ about the x-axis.

3.5.2 Precession of Equinoxes

The rotation axis of Earth undergoes precession due to torque exerted by the Sun and the Moon. It slowly revolves around the ecliptic pole at the rate of approximately $50''$ per year. The equatorial plane, which is perpendicular to this axis, also shifts due to this effect. In particular, the line of intersection of the equatorial and ecliptic planes, and hence the vernal equinox, also moves clockwise along the ecliptic plane at the same rate. The period of this precession is roughly 26,000 years. The rotation axis currently points roughly in the direction of the star Polaris. After 13,000 years its angle would change by roughly $47°$. Hence the night sky would look very different at that time. Furthermore, the current night sky is very different from what was seen in ancient times.

Due to this effect, the coordinates of astronomical sources change slowly in both the equatorial and ecliptic coordinate systems. The change in the ecliptic coordinates is easily obtained. The ecliptic plane and the pole are fixed. Hence the latitude β does not change. The azimuthal coordinate λ changes at the rate $d\lambda/dt \approx 50''$ per year. Using this, we can also obtain the rate at which the equatorial coordinates of an object change due to precession.

3.5.3 Equatorial Mounting of a Telescope

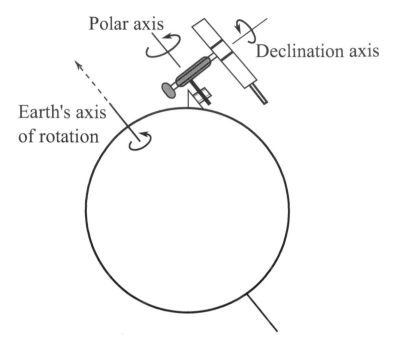

FIGURE 3.16: The two axes of rotation of a telescope in equatorial mount. One of the axes, the polar or the hour axis, is taken parallel to the Earth's axis of rotation. The declination axis is perpendicular to the hour axis.

In an equatorial mounting of a telescope one of the axes is taken to be parallel to the axis of rotation of the Earth. This is called the polar or the hour axis. The other axis, perpendicular to the hour axis, is called the declination axis (see Figure 3.16). While viewing any object in the sky it is necessary to keep the telescope pointing toward it for an extended period of time. In order to do so, one needs to continuously rotate the telescope to compensate for the Earth's rotation. In the equatorial mount, this is easily accomplished because we simply need to rotate about the hour axis at the same rate as the rotation of the Earth, but in the opposite direction. For all other mounts it is necessary to rotate the telescope simultaneously about both its axes.

We next describe how we can obtain the equatorial coordinates of an object using this mount. Because one of the axes is chosen to be the axis of rotation of Earth, the declination of an object can be directly obtained. It corresponds to the amount by which the telescope has to be rotated about the declination axis in order to point to the object and can be directly read off from the declination dial of the telescope. In order to obtain the RA, we first need to choose a reference for the azimuthal angle and locate the position of vernal equinox with respect to this reference. The reference is conveniently chosen to be the direction where the meridian intersects the equatorial plane toward the south, as shown in Figure 3.17. Let Γ denote the angle to the vernal equinox and h the hour angle of the source with respect to this reference. The hour angle corresponds to the rotation about the hour axis and is read off the hour angle dial of the telescope. Due to the rotation of Earth, both Γ and h increase at a steady rate. The right ascension α is given by

$$\Gamma = h + \alpha. \tag{3.17}$$

In practice it is convenient to determine the location of the vernal equinox by first finding the h of an easily recognizable star and using the catalogs to determine its α. The hour angle of the vernal equinox is then given by Equation 3.17.

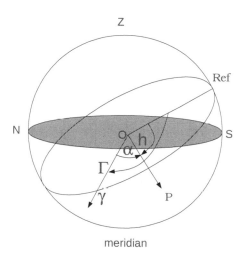

FIGURE 3.17: An illustration of equatorial mount. Here the observer is located at the center. The great circle NZS is the observer's meridian. The hour angle of a source P is measured clockwise from the reference point (Ref) on the celestial equatorial plane. Here γ shows the position of the vernal equinox. The angle between the vernal equinox and the source P is the right ascension α.

Exercises

3.1 The relationship between Cartesian and spherical polar coordinates is given in Equation 3.3. (a) Express x, y, z in terms of r and the equatorial coordinates (α, δ). (b) In an equatorial grid, the Cartesian coordinates of a star are $x = 10.0$, $y = 15.0$, $z = 6.0$ in some chosen units. Determine its distance r and the angular coordinates α and δ. (c) Repeat (b) for $x = -10.0$, $y = -15.0$, $z = 6.0$.

3.2 The proper motion of Barnard's star is $10.3''$ per year. It is located at a distance of 1.834 pc. Determine its transverse speed, v_t, in Km/s.

3.3 Consider a star that is moving away from us at a speed of 300 Km/s. It emits radiation of wavelength 500 nm. Determine the Doppler shift, $\Delta\lambda$, and the observed wavelength, λ.

3.4 Let $\vec{\mu}$ represent the proper or angular velocity of a star. We define $\mu_\delta = \dot{\delta} = d\delta/dt$ and $\mu_\alpha = \dot{\alpha} = d\alpha/dt$, where δ and α represent the Dec and RA of the star. Show that the components of the space velocity $\vec{v}$ are given by $v_\delta = r\mu_\delta$ and $v_\alpha = r\cos\delta\mu_\alpha$, where r is the distance of the star. Hence prove Equation 3.7. Hint: First consider a small angular displacement, $\Delta\delta$, along the longitude in time Δt. Find the distance traveled and hence determine the velocity. Repeat this for a small displacement along the latitude. Note that the two displacements can be considered independently.

3.5 Recall that at the summer solstice, the Sun never sets at latitudes close to the North Pole. (a) Find the range of latitudes for which this is true. (b) At any latitude l there exists a group of stars that always remain either above or below the horizon. These are called circumpolar stars. Find the range of declinations for these stars at latitude l. (c) Determine the names of a few bright circumpolar stars visible at your location. Locate them in the night sky and track their motion by observing them at different times.

3.6 Using the precession rate of $50''$ per year, verify that the period of precession is roughly 26,000 years.

3.7 Determine the inverse transformation from the ecliptic to equatorial coordinate system, that is, express the equatorial coordinates of a point in terms of its ecliptic coordinates.

3.8 Determine the rate at which the equatorial coordinates of a source change due to precession. Start by differentiating the equations obtained in the previous exercise with respect to time. This gives us $d\delta/dt$ and $d\alpha/dt$ in terms of β, λ, $d\lambda/dt$ and the transformation angle $\theta = 23°26'$. Next, eliminate β, λ in terms of δ and α to obtain the final result.

3.9 Determine the transformation between the equatorial and galactic coordinate systems following the procedure explained in the text. The galactic pole is located at $\delta = 27.13°$ and $\alpha = 192.86°$. The galactic center is located at $\delta = -28.94°$ and $\alpha = 266.40°$.

Chapter 4

Photometry

4.1 Introduction

We observe the stars and galaxies through the electromagnetic radiation they emit. We receive this radiation over a wide range of frequencies. In this chapter we will define terms such as intensity, flux density, and luminosity that characterize the observed radiation. The total amount of radiation emitted by a source per unit time is called its luminosity. This is a measure of the intrinsic brightness of the source. The units of luminosity are Joules per second (J/s), also called Watts (W). We denote the luminosity of any object with the symbol L. We are often interested in the radiation emitted in a narrow band of frequencies, centered at some frequency ν. This is characterized by the specific luminosity L_ν, which is defined as the energy emitted per unit time per unit frequency interval. The unit of frequency is Hz (Hertz). Hence the unit of L_ν is W/Hz.

The radiation that we receive from any astronomical object depends not only upon the luminosity of the source but also many other factors such as its distance, the attenuation of radiation during propagation, the area over which the radiation is received, etc. We observe this radiation with our eyes or with instruments such as telescopes, radio antennae, gamma ray detectors, etc. The human eye receives radiation over a relatively small area. Furthermore, it is sensitive only to the frequencies in the visible range. A telescope, in contrast, has a much wider aperture and hence collects more radiation in comparison to the eye. The amount of light or radiation an instrument receives also depends on the angle of observation. A telescope that points directly toward the source receives more light in comparison to one that points in a different direction. Furthermore, the amount of radiation an instrument detects depends on its sensitivity, which is a function of the frequency of radiation. It is useful to characterize the radiation field in terms of quantities such as flux and intensity, which are independent of these effects.

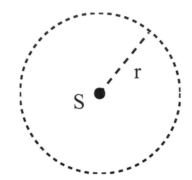

FIGURE 4.1: The radiation received from an isotropic source S is the same at all points at a distance r.

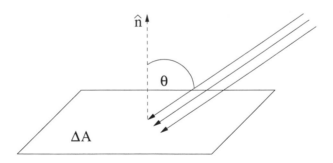

FIGURE 4.2: A beam of radiation, incident on a surface of area ΔA at an angle θ to the normal $\hat{n}$.

4.2 Flux Density and Intensity

Consider a source, such as the Sun, that emits radiation isotropically. Let us denote the luminosity of the Sun by L. Imagine an observer at a distance r from the Sun, as shown in Figure 4.1. At the position of the observer, the energy emitted by the Sun is distributed uniformly over the sphere of radius r. Hence the energy received by this observer, per unit area, is given by

$$F = \frac{L}{4\pi r^2}. \tag{4.1}$$

This is called the flux density. For an isotropic source, it decreases with distance as $1/r^2$. The units of flux density (F) are J/(s·m^2). The total flux passing through the spherical surface of radius r is equal to the luminosity L of the source.

In the above example we considered the flux passing through an area perpendicular to the incident solar radiation. This gives us the flux density at normal incidence. In general the area may be inclined at any angle to the incident radiation. Let us assume that radiative energy $\Delta \mathcal{E}$ is incident on a small surface of area ΔA in a small time interval Δt as shown in Figure 4.2. Let $\hat{n}$ be a unit normal to area A. The radiation is incident at an angle θ to $\hat{n}$. The flux density F is equal to the energy per unit area per unit time, that is,

$$F \approx \frac{\Delta \mathcal{E}}{\Delta A \cdot \Delta t}. \tag{4.2}$$

Here we have assumed that the radiation is uniform over a small time interval Δt and area ΔA. Eventually these intervals will be taken to be infinitesimally small. In this limit, the approximate equality in Equation 4.2 becomes exact.

We are often interested in the radiation field in a narrow range of frequencies. Let us assume that we receive the energy $\Delta \mathcal{E}$ in a small frequency band $\Delta \nu$ centered at frequency ν. The specific flux density F_ν is defined as the flux density per unit frequency interval, that is,

$$F_\nu \approx \frac{\Delta \mathcal{E}}{\Delta A \cdot \Delta t \cdot \Delta \nu}. \tag{4.3}$$

The units of F_ν are W/(m^2 Hz). Astronomers also use Janskys (Jy),

$$1\text{Jy} = 10^{-26} \frac{\text{W}}{\text{m}^2 \text{ Hz}} \tag{4.4}$$

This unit is particularly useful in radio astronomy because the flux density of astronomical radio sources is typically of this order of magnitude. The energy received at Earth from such sources is very small.

Let's consider a radio source with flux density equal to 1 Jy. Let's assume that we are making observations at 1 GHz using an antenna with an aperture of diameter 1 m^2. For simplicity we assume that the flux is constant over an interval of 1 GHz, centered at $\nu_0 = 1$ GHz, and zero outside this interval. We are interested in the total flux received by the antenna. The radiative energy collected by such an antenna in 1 day ($\approx 10^5$ s) is roughly equal to $10^{-26} \times 10^9 \times 10^5 = 10^{-12}$ Joules. The factor 10^9 arises because we are integrating the flux over a frequency interval of 1 GHz $= 10^9$ Hz $= 10^9$ 1/s. We can compare this with a typical macroscopic energy such as the energy required in lifting a paper. Let's assume that the paper has a mass of 1 gm and we need to lift it by 10 cm. The energy required is equal to the gravitational force mg times the distance h moved by the paper. Here m is the mass of the paper and the acceleration due to gravity, $g = 9.8$ m/s^2. The energy required in lifting the paper is therefore $mgh \approx 10^{-3}$ Joules. Hence the total energy collected by the antenna in a day is nine orders of magnitude smaller than this.

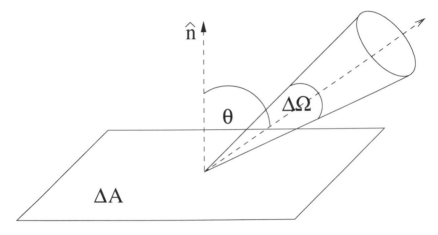

FIGURE 4.3: Radiation incident on a small area ΔA at an angle θ to normal, contained within a small solid angle $\Delta\Omega$.

In general, the radiation may be incident on a surface from several directions. For example, the sky constitutes a hemispherical source of radiation, in contrast to a star, which is almost a point object. Hence the radiation from the sky is incident upon any small area element at Earth from all possible directions in the upper hemisphere. We, therefore, define an observable, called the intensity, which contains information about the relative proportion of radiative energy incident from different directions. Let the radiative energy impinging on a small area element ΔA in time Δt from a direction θ relative to the normal, contained within a solid angle $\Delta\Omega$, be equal to $\Delta\mathcal{E}$ (see Figure 4.3). The concept of a solid angle is reviewed in Section 4.7. Furthermore, let the radiation observed lie in a narrow frequency range $[\nu, \nu + \Delta\nu]$. The specific intensity I_ν is defined as

$$I_\nu \approx \frac{\Delta \mathcal{E}}{\cos \theta \cdot \Delta A \cdot \Delta t \cdot \Delta \Omega \cdot \Delta \nu} , \qquad (4.5)$$

that is, it is the energy per unit time per unit area, projected normal to the angle of incidence, per unit solid angle and per unit frequency interval. The specific intensity, in general, depends on the time of observation, the direction of observation, and the frequency of radiation. It contains complete information about the strength of the radiation field at any given point.

The approximate equality in Equation 4.5 becomes exact when the small quantities, $\Delta \mathcal{E}$, ΔA, Δt, $\Delta \Omega$, and $\Delta \nu$, become infinitesimal. These are denoted by $d\mathcal{E}$, dA, dt, $d\Omega$, and $d\nu$, respectively. We obtain

$$I_\nu = \frac{d\mathcal{E}}{\cos \theta \cdot dA \cdot dt \cdot d\Omega \cdot d\nu} . \qquad (4.6)$$

The units of I_ν are

$$\frac{\text{Watts}}{\text{m}^2 \text{ Hz steradian}} .$$

The total intensity is defined as the integral of the specific intensity over all frequencies,

$$I = \int_0^\infty I_\nu d\nu . \qquad (4.7)$$

We also define the flux density F_ν at frequency ν by integrating over the radiation received from all directions, that is, the entire solid angle,

$$F_\nu = \int_{4\pi} I_\nu \cos \theta d\Omega = \int_0^{2\pi} d\phi \int_{-1}^{1} d\cos \theta I_\nu \cos \theta .$$

The total flux density is similarly related to the total intensity I by

$$F = \int I \cos \theta d\Omega .$$

The flux received by a surface of area A is defined as the radiative power passing through the surface

$$L_\nu = \int_A dA F_\nu ,$$

and similarly,

$$L = \int_A dA F .$$

L_ν and L have units of W/Hz and W respectively. The total flux or luminosity (L or L_ν) is defined as the flux passing through a closed surface encompassing a source.

We next consider some examples. The flux density of solar radiation at

Earth, integrated over all frequencies, is $F_E = 1,400$ J/(s·m^2). Because the Sun is an isotropic source, its luminosity is related to its flux density by the relationship

$$F_E = \frac{L_{Sun}}{4\pi R_{S-E}^2},$$

where R_{S-E} is the distance of the Sun from Earth. This yields the solar luminosity $L_{Sun} = 3.9 \times 10^{26}$ J/s. We next determine the intensity of the solar radiation at Earth. A person standing at Earth receives solar radiation only from the direction of the solar disk. Hence the intensity is non-zero only for a small range of directions corresponding to the solar disk. Let R_S be the radius of the Sun. The solid angle subtended by the Sun at any point on Earth is

$$\Delta\Omega \approx \frac{\text{cross} - \text{sectional area of Sun}}{(\text{Distance})^2} = \frac{\pi R_S^2}{R_{S-E}^2}.$$

This is valid in the limit $R_S << R_{S-E}$. The radiation is, therefore, received only within this solid angle. Consider an observer O at the surface of the Earth, as shown in Figure 4.4. The z-axis is normal to the surface at the position O. The radiation from the Sun is received at an angle θ_0 from normal. We use polar coordinates with the normal to the surface as the z-axis. The solar radiation is received along the z'-axis within a solid angle $\Delta\Omega$. Let $I_\nu(\theta, \phi)$ be the specific intensity of this radiation at any position (θ, ϕ), where the angular coordinates are measured with respect to the polar axis z. This angular position is denoted by the dashed line in Figure 4.4. The corresponding angular coordinates in terms of the polar axis z' are denoted as (θ', ϕ').

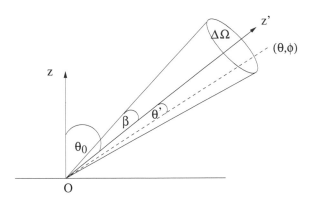

FIGURE 4.4: The solar radiation, received at an angle θ_0 with respect to the z-axis by an observer O within a solid angle $\Delta\Omega$ centered around the z'-axis. The z-axis is taken to be normal to the surface at the position of the observer O.

We can express $I_\nu(\theta, \phi)$ as

$$
\begin{aligned}
I_\nu(\theta, \phi) &= K_\nu \quad if \ \theta' \leq \beta, \\
&= 0 \quad if \ \theta' > \beta, \qquad (4.8)
\end{aligned}
$$

where $\beta \approx R_S/R_{S-E}$ is the angle subtended by the radius of the Sun. As we will see, K_ν is related to the flux density at the surface of the Sun. On the right-hand side of Equation 4.8 we have used the z'-axis to specify the boundary of the Sun. It is reasonable to assume that the radiation received at O from different points on the Sun is the same. In that case, K_ν depends only on the frequency but does not depend on the coordinates (θ', ϕ'). The total intensity can be obtained by integrating this equation over frequency. Because the radiation is contained within a solid angle $\Delta\Omega$, the specific flux density at frequency ν is approximately equal to $K_\nu \cos\theta_0 \Delta\Omega$. The factor $\cos\theta_0$ arises because the area on which the flux is being computed is oriented at an angle θ_0 with respect to the solar radiation. We can derive this formula for the flux density more rigorously by integrating I_ν over the solid angle subtended by the Sun. We obtain

$$
F_\nu = \int I_\nu \cos\theta d\Omega' = \int_0^\pi \sin\theta' d\theta' \int_0^{2\pi} d\phi' I_\nu \cos\theta
$$

$$
= \int_{-1}^1 d\cos\theta' \int_0^{2\pi} d\phi' I_\nu \cos\theta.
$$

The integral is complicated because $\cos\theta$ has a complicated dependence on θ' and ϕ'. However, we can simplify it by making the approximation

$$
\cos\theta \approx \cos\theta_0 = \text{const}.
$$

This follows because I_ν is non-zero only over a small range of values of the polar angle θ centered around θ_0. We can pull the $\cos\theta_0$ factor out of the integral over the angular variables. The resulting integral over the angular coordinates simply gives the solid angle subtended by the Sun. We, therefore, find

$$
F_\nu \approx K_\nu \cos\theta_0 \Delta\Omega. \qquad (4.9)
$$

We next relate this to the solar radiant energy received at Earth. The solar luminosity, integrated over frequencies, is $L_{Sun} = 3.9 \times 10^{33}$ ergs/s. The integrated solar flux density, at normal incidence $(\theta_0 = 0)$, at the surface of the Earth is $F_E = L_{Sun}/(4\pi R_{S-E}^2) = 1,400$ J/(s· m^2). This is related to the flux given in Equation 4.9, integrated over frequencies. The only quantity in Equation 4.9, which depends on frequency is K_ν. Hence we obtain

$$
F_E = \Delta\Omega \int_0^\infty K_\nu d\nu = K\Delta\Omega.
$$

Here K is the integral of K_ν over all frequencies. This implies that

$$K = \frac{F_E}{\Delta\Omega} = \frac{L_{Sun}}{4\pi^2 R_S^2} = \frac{F_S}{\pi},$$

where F_S is the flux density at normal incidence at the surface of the Sun. Hence we find that the factor K (or K_ν) and hence the intensity is a measure of the flux density at the surface of the Sun and is independent of the distance of the observer from the source.

The independence of intensity with respect to distance is a general property. We can understand this by considering a simple example. Consider an isotropic source depicted in Figure 4.1. The flux density at a distance r decreases as $1/r^2$. The intensity is equal to flux density per unit solid angle. The solid angle subtended by a detector of fixed area also decreases as $1/r^2$. The two $1/r^2$ factors cancel and we find that the intensity is independent of r

We next consider the radiation received from a star. The main difference is that a star subtends a very small solid angle. We can continue to use the same formulae as for the Sun, but with the value of $\Delta\Omega$ taken to be extremely small. Using Equation 4.9, we find $F_\nu = K_\nu \Delta\Omega \cos\theta_*$, where $\Delta\Omega = \pi R_*^2/r^2$ is the solid angle subtended by the star, R_* is its radius and θ_* its polar angle, as in the case of Sun.

As an example, we compute the flux density of the brightest star, Sirius, at Earth. Its luminosity is roughly 25 times that of the Sun and its distance from Earth is 2.6 pc. Hence its flux density F at normal incidence, is 1.3×10^{-7} J/(s· m^2).

4.3 Blackbody Radiation

All objects emit radiation that is characterized by their temperature. For example, the Sun, which has a surface temperature of 5,770K (Kelvin), emits radiation at visible frequencies. Similarly, many objects glow, that is, emit visible radiation, when heated to high temperatures. In fact, all objects emit radiation as long as their temperature is different from absolute zero (0K or $-273°$C). For example, the walls of our room radiate energy that is characterized by the room temperature ($\sim$300K)[1]. The radiation emitted by an object at this temperature, however, is very different from that emitted by hotter objects, such as the Sun. This radiation cannot be observed by the human eye because it is emitted predominantly at infrared frequencies. The mean frequency of emitted radiation decreases with the temperature of the object.

We next consider radiation that is incident upon an object. This radiation

---

[1]The notation $\sim$ means roughly equal to.

is partially absorbed, partially reflected, and partially transmitted by the object. The precise fraction that is absorbed, reflected, or transmitted depends on the nature of the object and the frequency of the incident wave. For example, a piece of wood does not allow transmission of visible light. In contrast, glass allows almost all the light to be transmitted. However, glass does not allow free transmission at all frequencies. It is transparent at visible frequencies but becomes opaque at lower (infrared) and higher (ultraviolet) frequencies. Another very interesting example is the Earth's atmosphere, which is almost perfectly transparent to visible radiation but becomes highly opaque as we move to higher and lower frequencies.

Besides transmission, an object also reflects and absorbs radiation incident upon it. Hence as the solar radiation falls upon Earth, part of it is absorbed and the remainder reflected back. The reflected radiation has the same frequency as the incident radiation. It is due to this reflected solar radiation, which lies at visible frequencies, that we are able to see objects during the daytime. The part that is absorbed by Earth is later emitted at infrared frequencies. Notice that the Earth-Sun system is at steady state. The mean temperature of the Sun as well as the Earth does not change appreciably with time. Hence the radiation that Earth absorbs must be equal to the radiation it emits. Or else its mean temperature would change.

In general, the amount of radiation an object absorbs depends on the object as well as the frequency of the incident radiation. A blackbody is defined as an ideal object that absorbs all the radiation incident upon itself. The term *blackbody* is well suited because a body appears black to us if it absorbs all the visible light incident upon itself. However, the term *blackbody* is used here in a more general sense. An ideal blackbody absorbs radiation of all frequencies, not just the visible light. None of the incident radiation is reflected or transmitted. The blackbody also emits radiation, called blackbody radiation, with a characteristic spectrum called the blackbody spectrum. The spectrum is fixed by a single parameter, the temperature of the object. The specific intensity B_ν of a blackbody radiation emitted by a body at temperature T (in Kelvin) can be expressed as

$$B_\nu(T) = \frac{2h\nu^3/c^2}{e^{h\nu/kT} - 1}, \qquad (4.10)$$

where ν is the frequency, h is Planck's constant, c is the velocity of light, and k is the Boltzmann constant. We can obtain the specific intensity in terms of the wavelength $\lambda = c/\nu$ by making a change in variables. We obtain

$$B_\lambda = \frac{2hc^2}{\lambda^5} \frac{1}{e^{hc/kT\lambda} - 1}. \qquad (4.11)$$

The intensity B_λ is defined such that

$$\int_0^\infty d\lambda B_\lambda = \int_0^\infty d\nu B_\nu = B. \qquad (4.12)$$

In Figure 4.5 we show the wavelength dependence of specific intensity for two different temperatures. The peak position of the blackbody spectrum is given by Wien's displacement law, which states that

$$\lambda_{max}T = 0.290 \text{ cm K}, \tag{4.13}$$

where λ_{max} is the wavelength at which B_λ is maximum. The total intensity of blackbody radiation is obtained by integrating the specific intensity over all frequencies. We find

$$B = \sigma T^4/\pi, \tag{4.14}$$

where

$$\sigma = 5.670 \times 10^{-5} \; \frac{\text{erg}}{\text{s cm}^2 \text{ K}^4}, \tag{4.15}$$

is the Stefan–Boltzmann constant. The total luminosity of a blackbody is

$$L = \sigma T^4 S, \tag{4.16}$$

where S is its surface area.

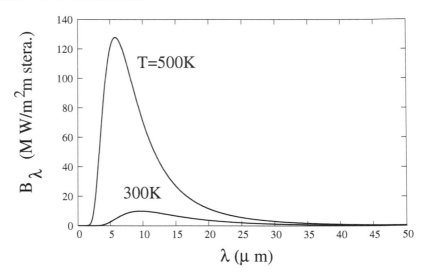

FIGURE 4.5: The wavelength dependence of the blackbody specific intensity, B_λ, in units of Mega Watts/(m^2·m·steradian) corresponding to temperatures, T=300K, 500K.

An ideal blackbody emits the maximum energy possible for an object at a particular temperature. All other objects at the same temperature have lower emission. We can account for this difference by introducing an efficiency factor ϵ in the formula for total luminosity. Hence the luminosity of a real body can be expressed as

$$L = \epsilon\sigma T^4 S, \tag{4.17}$$

where the factor ϵ lies in the range 0 to 1. An ideal blackbody can be constructed by making a cavity whose walls are opaque. There is a small hole through which radiation can enter the cavity. This radiation interacts with the walls of the cavity and undergoes repeated reflections. It also gets partially absorbed and re-emitted by the walls. However, because the walls are opaque, it will never leave the cavity, which, therefore, acts as a perfect absorber. The walls eventually acquire thermal equilibrium at a certain temperature T. The radiation in the cavity now reaches a steady state with no further exchange of energy with the cavity walls. We will also assume that the cavity is sufficiently large so that the geometry of the walls does not affect the nature of the radiation field. This can of course be true only if we are not too close to the cavity walls. The radiation in such a cavity is almost the same as an ideal blackbody radiation. This radiation can be sampled through a small hole in the cavity walls.

The radiation inside the cavity is isotropic because the intensity does not depend on the direction of observation. At any point the intensity of radiation is the same in all directions. The intensity in any direction is exactly equal to that in the opposite direction. Hence the flux density through any area element is zero. We can explicitly verify this using the formula for the flux density,

$$F_\nu = \int_0^\pi \sin\theta d\theta \int_0^{2\pi} d\phi B_\nu \cos\theta = 0\,.$$

The integral over θ gives zero because B_ν is independent of θ. In general, the flux density of an isotropic radiation field is zero.

It is important to understand that the formula for the specific intensity of a blackbody radiation, Equation 4.10, is really valid only inside the cavity. This formula says that the intensity is independent of direction. In other words, an observer would see exactly the same intensity in all directions. This is clearly not valid for an observer outside the cavity. Such an observer would receive radiation only from the direction of the small hole and nothing from other directions. Hence the radiation is isotropic only inside the cavity and not outside. However, the frequency dependence of the radiation sampled through the hole is correctly given by Equation 4.10.

Blackbody radiation plays a central role in astrophysics. The radiation emitted by stars is well approximated by the blackbody spectrum. A typical star is spherical symmetric to a very good approximation. Hence it is reasonable to assume that the radiation it emits is independent of the angular position of the observer with respect to the center of a star. The interiors of stars are well described by assuming thermal equilibrium. The temperature of a star slowly decreases with distance from the center. The intensity of radiation at any point inside the star at temperature T is well approximated by the blackbody formula corresponding to the temperature at that location. At the surface of the star, however, the conditions of thermal equilibrium are not satisfied. At this point we have free emission. In contrast to the interior, the radiation is no longer isotropic. An observer at the surface will receive radiation only from half the solid angle, that is, from the interior of the star. The radiation from the exterior is negligible because it arises only due to the sparse stellar atmospheres. The spectral dependence of the radiation received by the observer, however, is correctly described by the blackbody formula.

The concept of blackbody radiation also plays a central role in cosmology. As we will discuss in Chapter 16, the entire Universe is immersed in a radiation field, called Cosmic Microwave Background Radiation (CMBR). The temperature of this radiation is 2.73K. It turns out that this is the best blackbody spectrum seen to date. The experimentally observed spectrum matches the theoretical prediction, Equation 4.5, to one part in 10,000.

4.4   Energy Density in an Isotropic Radiation Field

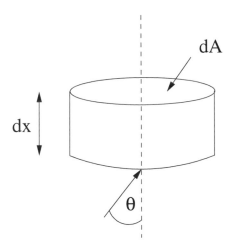

FIGURE 4.6: An infinitesimal cylindrical volume element dV of length dx and base area dA.

Consider an isotropic radiation field, such as the blackbody radiation. We are interested in determining the energy density, that is, energy per unit volume, of this field at any point. Let the specific intensity of radiation at any point at frequency ν be I_ν. We denote the corresponding energy density by U_ν. Consider an infinitesimal volume element dV in this radiation field, as shown in Figure 4.6. Let the cross-sectional area of the volume element be dA and length dx. Let dE_ν be the radiation contained in this volume in the frequency interval $d\nu$. Then the energy density of the radiation field at frequency ν is

$$U_\nu = \frac{dE_\nu}{dV\,d\nu}. \tag{4.18}$$

In order to compute dE_ν, we consider radiation incident on the walls of the volume element. We will eventually take the limit, $dx \to 0$, $dA \to 0$. For convenience, the limiting process is done such that we first take the limit $dx \to 0$ and then $dA \to 0$. In this case we only need to consider the radiation incident on the upper and lower surfaces. Furthermore, for sufficiently small dx, all radiation striking the lower surface will leave from the upper surface. The distance traveled by radiation incident at angle θ to the normal at the lower surface inside the volume element is $dx/\cos\theta$. Hence the time it spends inside this volume element is $dt = dx/(c\cos\theta)$. The energy contained inside dV due to radiation incident on lower surface is

$$dE_\nu^1 = \int_\Omega I_\nu \cos\theta \, dA d\nu d\Omega dt \,,$$

where the integral is over half the solid angle because only the radiation propagating upward is incident on the lower surface. Hence we find

$$dE_\nu^1 = \int_\Omega I_\nu dA d\nu d\Omega dx/c = 2\pi I_\nu dV d\nu/c \,,$$

where we have used $dV = dAdx$ and the fact that I_ν is independent of angles, that is, the radiation is isotropic. We get an identical contribution dE_ν^2 due to radiation striking the upper surface. Adding the two, we obtain

$$dE_\nu = 4\pi I_\nu dV d\nu/c \,, \tag{4.19}$$

which gives

$$U_\nu = \frac{4\pi I_\nu}{c} \,. \tag{4.20}$$

Integrating over all frequencies, we obtain the total energy density $U = 4\pi I/c$. For the blackbody radiation, the intensity is given by Equation 4.14. Hence we obtain

$$U = \frac{4\sigma T^4}{c} \,. \tag{4.21}$$

4.5 Magnitude Scale

In astronomy we use a logarithmic scale to specify the flux density and the luminosity of astronomical objects. This system originated when early observations were done with the naked eye, whose response is close to logarithmic. Although most observations are now made by instruments, this system is still in use.

4.5.1   Apparent Magnitude

The flux density F received at Earth from a star can be expressed in terms of the apparent magnitude, m, defined as

$$m = -2.5 \log_{10} \frac{F}{F_0} \,,$$

where F_0 is a reference flux density. The constant 2.5 is used for historical reasons. For two stars with flux densities F_1 and F_2,

$$m_1 - m_2 = -2.5 \log_{10} \frac{F_1}{F_2} \,.$$

The star with larger m appears less bright. If $m_1 - m_2 = 5$, then $F_2 = 100F_1$.

The magnitude system is designed to specify the flux of an object with respect to the flux of some standard source used as a reference. This is convenient because measurement of relative flux is often much easier and more reliable in comparison with the absolute flux. Many instrumental errors as well as distortions due to atmosphere cancel out when measuring the ratio of two fluxes but will lead to errors in measurement of an absolute flux. The magnitude of the reference source is assigned some fixed value, such as zero, as discussed below.

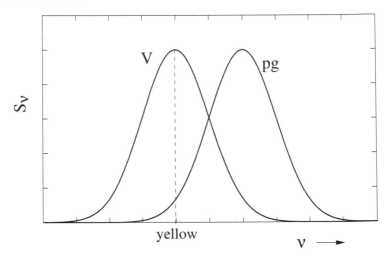

FIGURE 4.7: An illustration of the sensitivity function, S_ν, of the human eye (visual V) and the photographic plate (pg) as a function of the frequency, ν. The sensitivity of eye peaks at $\lambda \approx 550$ nm corresponding to yellow color.

The flux and hence the observed magnitude depend on the instrument, which may have different sensitivities at different wavelengths. Hence, what is measured is

$$F = \int_0^\infty S_\nu F_\nu d\nu \,, \tag{4.22}$$

where S_ν is the sensitivity function of the instrument and ν is the frequency. For example, the human eye is most sensitive to yellow as shown in Figure 4.7. Its sensitivity decreases at higher and lower frequencies. For comparison, we also show the sensitivity of a typical photographic plate in Figure 4.7. The photographic plate is typically more sensitive to higher frequencies in comparison to the human eye. It is now obsolete but historically played a very important role in astronomy. We now make more accurate measurements using photometers. Furthermore, standardized broad band filters, centered at different frequencies, are used. Some of the standard filters are U, B, V, and R, which refer to ultraviolet, blue, visual, and red wavelengths, respectively. The apparent magnitudes of these filters are denoted by m_U, m_B, m_V, and m_R, respectively. These are also denoted simply as U, B, V, and R, respectively. The effective wavelengths for the U, B, V, and R filters are 365 nm, 445 nm, 551 nm, and 658 nm, respectively. Their full widths at half maximum are 66 nm, 94 nm, 88 nm, and 138 nm, respectively. The visual magnitude m_V corresponds closely to the sensitivity of the human eye.

There exist several different magnitude systems in astronomy corresponding to different choices of the reference flux. One standard system uses the star Vega as a reference and assigns a value equal to 0.03 for all its apparent magnitudes, such as U, B, V, and R.

4.5.2 Absolute Magnitude

The apparent magnitude is related to the flux density of an astronomical object. We also need a measure that quantifies the luminosity or intrinsic brightness of an object. This is provided by the absolute magnitude, denoted by M. It is defined as the apparent magnitude of an object when it is placed at a distance of 10 pc from the observer. For a source that radiates isotropically, the flux density $F(r)$ in vacuum is proportional to $1/r^2$, where r is the distance of the observer from the source. Hence,

$$\frac{F(r)}{F(10)} = \left(\frac{10 \text{ pc}}{r}\right)^2 , \qquad (4.23)$$

where $F(10)$ is the flux density at a distance of 10 pc. This implies that

$$m - M = -2.5 \log_{10} \frac{F(r)}{F(10)} = -2.5 \log_{10} \left(\frac{10 \text{ pc}}{r}\right)^2 = 5 \log_{10} \frac{r}{10 \text{ pc}} . \qquad (4.24)$$

The difference $m - M$ is called the distance modulus because it is a measure of the distance of the object. The absolute magnitudes corresponding to filters U, B, V, and R are denoted by M_U, M_B, M_V, and M_R, respectively. For example, the absolute visual magnitude, M_V, of the Sun is 4.83. In Equation 4.24, we have ignored the extinction of light due to propagation. This will be discussed in Chapter 7.

The measurement of absolute magnitude is clearly much more complicated in comparison to that of apparent magnitude. It requires knowledge of the distance from the source. A direct measurement of distance is possible only using parallax. This can be used only for nearby sources. For larger distances, the luminosity is deduced indirectly. One often uses the fact that the luminosity of many sources shows some definite relationship to some other observables. In some cases, the luminosity is approximately constant for all the sources belonging to a particular class. Hence one is able to indirectly deduce the absolute magnitude of these sources.

4.5.3   The Color Index

The color index is defined as the difference in the apparent magnitudes corresponding to two different filters. Thus, $U - B$, $B - V$ are examples of color indices. It is a measure of the color of an object in comparison to a reference star such as Vega. Because all the apparent magnitudes of Vega are chosen to be 0.03, all its color indices are equal to zero. Hence, if for some star $U - B$ is positive, this means that the object radiates more strongly in blue in comparison to ultraviolet relative to the reference star Vega. Using Equation 4.24, we find that we can also express a color index as the difference in absolute magnitudes. In particular, we find that

$$U - B = M_U - M_B \,, \tag{4.25}$$

with similar equations for other color indices.

4.5.4   Bolometric Magnitude

The apparent magnitude is a measure of the flux density received from a star within a certain wavelength range. It is also useful to have a measure of the total flux, integrated over all wavelengths. The magnitude corresponding to this flux is called the bolometric magnitude. It is of course impossible to make a measurement of the flux over all the wavelengths. In most cases, however, the flux dies off rapidly beyond a certain wavelength regime. Hence a limited measurement provides an excellent approximation of the bolometric magnitude. The spectrum of many stars peaks at visible or ultraviolet wavelengths. Their luminosity dies off rapidly as we go to higher frequencies in the ultraviolet or lower frequencies in the infrared regime. Hence a measurement covering the infrared, visible and ultraviolet frequencies provides a good estimate of the bolometric magnitude of these objects.

4.6 Stellar Temperatures

An important parameter is the temperature at the surface of the star. The surface temperature determines the nature of radiation, the continuous spectra, the discrete lines and the bands, received from the star. The temperature is deduced from the observed radiation. Astronomers define several different temperatures, which depend on some particular property of the radiation received. These rely on the fact that the continuum radiation emitted by stars approximately follows the blackbody spectrum. Hence one tries to deduce the corresponding blackbody temperature from observations.

4.6.1 Effective Temperature

The effective temperature T_e is defined as the temperature of a blackbody that radiates the same total flux as the star. Consider a star of radius R located at a distance r from Earth. The flux density at the surface of the star is $F_S = \sigma T_e^4$, which implies that the luminosity of the star is $L = 4\pi R^2 F_S$. The flux density F_E at Earth is given by

$$F_E = \frac{L}{4\pi r^2} = \frac{R^2}{r^2} F_S = \frac{\sigma \delta^2 T_e^4}{4},$$

where $\delta = 2R/r$, is its angular diameter. Hence an observation of the flux density F_E and the angular diameter gives an estimate of T_e.

All stars, besides the Sun, appear as point-like objects. A direct measurement of their angular diameter is very difficult. For some stars, angular diameters have been determined using interferometry or lunar occultations. Furthermore, diameters can be determined under some special circumstances for stars in binary systems.

4.6.2 Color Temperature

We can also extract the temperature of a star by measuring its apparent magnitude corresponding to two different filters. Let $F_E(\nu_1)$ and $F_E(\nu_2)$ be the flux densities of a star observed at Earth at two different frequencies, ν_1 and ν_2, respectively. The ratio of the flux densities, assuming that the radiation emitted follows the blackbody spectrum, is given by

$$\frac{F_E(\nu_1)}{F_E(\nu_2)} = \frac{B_{\nu_1}(T)}{B_{\nu_2}(T)},$$

where $B_\nu(T)$ is the intensity of a blackbody at temperature T. This relation provides an estimate of temperature, given a measurement of $F_E(\nu_1)$ and $F_E(\nu_2)$. In practice, one measures the difference in magnitudes corresponding to two broad band filters, for example $B - V$. The difference in magnitudes at two different frequencies can be expressed as

$$m_1 - m_2 = -2.5 \log_{10} \frac{F_E(\nu_1)}{F_E(\nu_2)}.$$

In the case of a broad band filter we require the integrated flux density corresponding to that filter. Hence the flux densities are computed by using Eq. 4.22 with F_ν equal to the blackbody intensity and S_ν the sensitivity function of the filter. The temperature extracted by this procedure is called the color temperature, T_c. Hence the color indices, such as $B - V$, provide a measure of the color temperature of a star.

If a star behaves as a perfect blackbody, then the color temperature would be the same as the effective temperature. However, in general, stars are not ideal blackbodies and these temperatures differ from one another. The color temperature is easiest to extract because it only requires measurement of apparent magnitudes. Hence the early estimates of temperature, made toward the beginning of the twentieth century, used this measure. The effective temperature can be deduced by spectroscopic measurements. The relationship between the effective temperature and the color index, $B - V$, is shown in Figure 4.8.

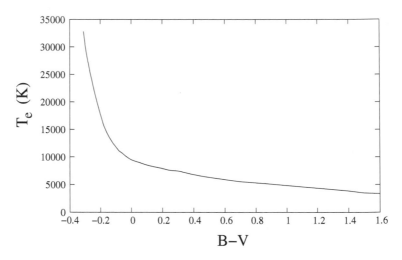

FIGURE 4.8: The effective temperature T_e of stars corresponding to different values of $B - V$. The error in effective temperatures ranges roughly from ± 2 % to ± 5 %. The actual values also depend on other parameters besides $B - V$, such as the metallicity of a star. This dependence is not shown in the above plot. (Data taken from E. Böhm-Vitense, *Annual Review of Astronomy & Astrophysics*, 19, 295 (1981).)

4.7 Appendix: Solid Angle

Consider a spherical shell of radius r, as shown in Figure 4.9. The solid angle Ω that a segment of the shell of area S subtends at the center of the sphere is given by

$$\Omega = \frac{S}{r^2}.$$

The unit of the solid angle is steradians. As a simple example we can determine the solid angle subtended by half the spherical shell. The area of this shell is $2\pi r^2$. Hence the corresponding solid angle is 2π steradians. The solid angle subtended by the full sphere is 4π steradians.

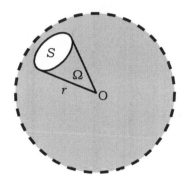

FIGURE 4.9: The section S of a spherical shell, with center located at O and radius r, subtends a solid angle Ω at O.

We are often interested in the solid angle subtended by a small area at some point. Consider the situation shown in Figure 4.10, where we have a small flat surface of area A. We are interested in the solid angle it subtends at point O. This solid angle depends on the orientation of the area element. Consider the line OP joining O to the point P, located at the center of the area element. If this line is perpendicular to the area, then the solid angle subtended is given approximately by

$$\Omega \approx \frac{A}{r^2},$$

where r is the distance of the area element from O. This is an approximate relation because strictly speaking, the area element should be part of a spherical shell, as shown in Figure 4.9. However, for small area elements, the area of the curved surface is approximately the same as that of the flat surface.

In general, the line OP may not meet the area at a right angle. Let us assume that the angle that the normal, $\hat{n}$, to the surface makes with OP is θ, as shown in Figure 4.10b. In this case, the solid angle is given approximately by

$$\Omega \approx \frac{A \cos \theta}{r^2},$$

that is, the solid angle subtended by the projected area element normal to OP.

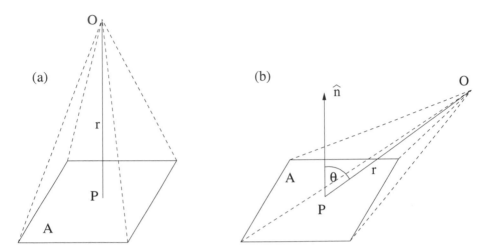

FIGURE 4.10: (a) A small flat surface of surface area A subtends a solid angle, approximately equal to A/r^2 at O. Here OP is perpendicular to the surface and its length is equal to r. The area A is assumed to be very small. (b) Here $\hat{n}$ denotes a normal to the flat surface A and the line OP makes an angle θ to normal. The solid angle subtended by a small area A at O is approximately equal to $A \cos \theta / r^2$.

Exercises

4.1 The unit of energy is the Joule (J) or erg.

$$1 \text{ J} = 1 \text{ Kg} \cdot \text{m}^2/\text{s}^2, \qquad 1 \text{ erg} = 1 \text{ g} \cdot \text{cm}^2/\text{s}^2.$$

Clearly, 1 J = 10^7 ergs. (a) The kinetic energy (K.E.) of a particle of mass M moving at speed v is equal to $Mv^2/2$. Determine the kinetic energy of a person of mass 60 Kg moving at speed 4 Km/hour. (b) Heat is also a form of energy. It is caused by the random motions of atoms and molecules of an object. As we heat an object, the kinetic energies of these particles increase. The heat capacity of water is 4,180 J/Kg·K or 4,180 Joules per kilogram per degree Centigrade. This means that to increase the temperature of 1 Kg of water by 1° Centigrade (C), we need to supply 4,180 Joules of energy. Find the energy required to heat 1 Kg of water at 30°C (room temperature) to 100°C.

4.2 According to the special theory of relativity, a particle of mass m at rest has a rest mass energy

$$E = mc^2. \tag{4.26}$$

Determine the rest mass energy of a person of mass 60 Kg. Compare with the kinetic energy, determined in the Exercise 4.1a.

4.3 Consider the diffuse radiation from the sky. Assume that it is uniform over the entire sky. Mathematically express the intensity received at the surface of the Earth. You can assume that roughly 20% of the total radiative energy received from the Sun, above the Earth's atmosphere, reaches the surface in the form of diffuse radiation.

4.4 The luminosity of the Sun is 3.839×10^{26} W. Its apparent magnitude is $m_{\text{Sun}} = -26.81$. Find its absolute magnitude and the distance modulus.

4.5 Verify the formula for B_λ given in Equation 4.11.

4.6 Integrate the formula for blackbody specific intensity, B_ν, in order to obtain an expression for the Stefan–Boltzmann constant. You can use the integral

$$\int_0^\infty dx \frac{x^3}{e^x - 1} = \frac{\pi^4}{15}. \tag{4.27}$$

4.7 Show that the radiation flux at the surface of a star is $F = \sigma T^4$. Hence show that its total luminosity is given by Equation 4.16.

4.8 The surface temperature of the Sun is 5770K. Write down the formula for the specific intensity, I_ν, of the solar radiation received by an observer at Earth. You may assume that the Sun is an ideal blackbody. You must carefully specify both the frequency as well as angular dependence of I_ν. Compute the solar flux at Earth and compare with the measured value 1,400 W/m^2.

4.9 Find the peak wavelength and the corresponding frequency of solar radiation, given that it is a blackbody at temperature $T = 5770$K.

4.10 Make a rough estimate of Earth's mean temperature T by assuming that it is a perfect blackbody. Assume that it absorbs all the solar radiation incident upon it and radiates it as a blackbody at temperature T. Assume that Earth is spherically symmetric. Numerically estimate the value of T.

4.11 Find the peak position of wavelength and the corresponding frequency for the CMBR.

Chapter 5

Gravitation and Kepler's Laws

The gravitational force plays the most prominent role in astrophysics. It is responsible for holding the solar system, the galaxies, and the galaxy clusters together. The gravitational force is well described by the Newton's inverse square law. Consider two point particles of mass M and m. Here we treat heavenly bodies, such as Sun, Earth, etc., as particles. Let $\vec{r}$ denote the position vector of m with respect to M, as shown in Figure 5.1. The force on m due to M is given by

$$\vec{F} = -G\frac{Mm\hat{r}}{r^2}, \tag{5.1}$$

where $\hat{r}$ is the unit vector along $\vec{r}$. The negative sign indicates that the force is attractive, that is, toward M. This law is applicable as long as the particles move at speeds much smaller than the speed of light. For most astrophysical applications, Newton's law is sufficiently accurate. The gravitational force is a central force. This means that its magnitude is a function only of the distance r, and the direction is along the displacement vector $\vec{r}$.

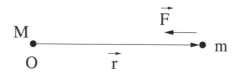

FIGURE 5.1: The point masses, M and m, are located at the origin O and at the position vector $\vec{r}$, respectively. The gravitational force $\vec{F}$ on m due to M points toward the origin.

It is also convenient to define the concept of a gravitational field. The gravitational field at any point is equal to the gravitational force on a point particle of unit mass located at that point. Hence if $\vec{g}(\vec{r})$ is the gravitational field at position $\vec{r}$, then the force on a particle of mass m at that point is

$$\vec{F} = \vec{g}m. \tag{5.2}$$

The gravitational field of a particle of mass M, at a position $\vec{r}$, is

$$\vec{g} = -G\frac{M\hat{r}}{r^2}. \tag{5.3}$$

5.1 Two-Body Problem

Consider the motion of two particles under the influence of their mutual gravitational force. We are interested in predicting their trajectories. This problem can be solved analytically. Let us first consider the simple case where one of the masses is very heavy. In that case we can assume that the heavy mass is at rest and focus on the motion of the light particle. This is applicable, for example, for the Sun-Earth and the Earth-Moon systems. Let us denote the light mass by m and the heavy mass by M. We choose the origin at the position of the heavy particle (see Figure 5.1). The gravitational force on m points toward O. Hence the torque acting on m, with respect to the origin, is zero and its angular momentum remains constant with time. The angular momentum, $\vec{L}$, of m is $\vec{L} = m\vec{r} \times \vec{v}$, where $\vec{v}$ is its velocity. Because $\vec{L}$ remains fixed and is perpendicular to $\vec{r}$, $\vec{r}$ must lie in a plane. This means that the trajectory of m lies in a plane. We take this to be the $x - y$ plane. The precise nature of the trajectory depends on the total mechanical energy, E, of m. We can express E as

$$E = K + U = \frac{1}{2}mv^2 + U, \tag{5.4}$$

where K is the kinetic energy of m, $v = |\vec{v}|$ its speed, and U its potential energy. It is given by

$$U = -\frac{\alpha}{r}, \tag{5.5}$$

where $\alpha = GMm$. It is convenient to use plane polar coordinates (r, θ) to analyze the motion of m. The velocity vector in these coordinates is given by

$$\vec{v} = \dot{r}\hat{r} + r\dot{\theta}\hat{\theta}, \tag{5.6}$$

and hence the kinetic energy K is

$$K = \frac{1}{2}m\dot{r}^2 + \frac{1}{2}(r\dot{\theta})^2 = \frac{1}{2}m\dot{r}^2 + \frac{L^2}{2mr^2}, \tag{5.7}$$

where $L = |\vec{L}|$ is the magnitude of the angular momentum vector. The total energy E and the angular momentum L are constants of motion and have to be specified as initial conditions.

We now distinguish three different cases:

1. $E > 0$: In this case, the mass m is not bound to M. The trajectory is a hyperbola, as shown in Figure 5.2. In plane polar coordinates, the solution can be written as

$$r = \frac{r_0}{1 - \epsilon \cos \theta},$$ (5.8)

where $r_0 = L^2/(\alpha m)$ and

$$\epsilon = \sqrt{1 + \frac{2EL^2}{\alpha^2 m}} > 1.$$

The particle comes in from very large distances, approaches the mass M, and then moves away.

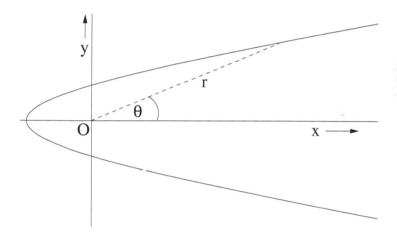

FIGURE 5.2: The hyperbolic orbit of a particle moving in the gravitational field of a very heavy particle located at the origin O. Here r and θ are the plane polar coordinates.

2. $E < 0$: In this case, the mass m is bound to M. The trajectory is an ellipse (see Figure 5.3) and satisfies the same equation, Equation 5.8. In this case, $E < 0$ and hence $0 \le \epsilon < 1$. The motion of the particle is periodic with the time period T given by

$$T^2 = \frac{4\pi^2 a^3}{(M + m)G} \approx \frac{4\pi^2 a^3}{MG}.$$ (5.9)

where a is the semi-major axis of the ellipse. At $\theta = 0$, $r = r_{max} = r_0/(1 - \epsilon)$ is maximum. At $\theta = \pi$, the particle reaches its minimum distance, $r = r_{min} = r_0/(1 + \epsilon)$ from the origin. After a time interval T, it returns to r_{max}. The speed of the particle depends on time. It is smallest at $r = r_{max}$ and largest at $r = r_{min}$. The parameter ϵ is equal to the eccentricity of the ellipse. It can be expressed as

$$\epsilon = \frac{r_{max} - r_{min}}{r_{max} + r_{min}}.$$
(5.10)

For ϵ close to 1, the ellipse is highly elongated. For $\epsilon = 0$, the ellipse reduces to a circle.

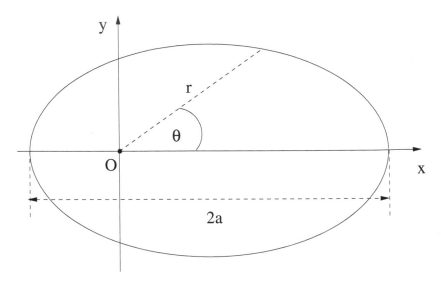

FIGURE 5.3: The elliptical orbit of a particle moving in the gravitational field of a very heavy particle located at the origin O.

3. $E = 0$: The mass m is barely unbound. The trajectory is a parabola, with $\epsilon = 1$, $r = r_0/(1 - \cos\theta)$. Let $\vec{v}_E$ denote the velocity of the particle at a distance r. This is called the escape velocity because the particle can escape the gravitation pull of M if its speed $v \geq v_E$. Using Equation 5.4, we obtain

$$v_E = \sqrt{\frac{2GM}{r}}.$$
(5.11)

Let us next consider the case of two particles without making the assumption that one of them is very heavy. In this case, both particles move about their common center of mass. This is illustrated in Figure 5.4. We are interested only in the relative motion of the two particles. Choosing M as the origin, the two-body problem reduces to a one-body problem with the energy again given by Equation 5.4, if we replace the mass m in the kinetic energy by the reduced mass μ, defined as

$$\mu = \frac{Mm}{M + m}.$$
(5.12)

The solution is again given by Equation 5.8, with m in r_0 and ϵ replaced by μ. Hence if the two particles are bound to one another, then mass m would move along an elliptical trajectory with M at the origin O, as shown in Figure 5.3. Similarly, the mass M will also appear to move in an elliptical orbit, as seen by an observer at m. For an observer located at the center of mass C, the position vector of m is given by

$$\vec{r}_2 = \frac{m}{M+m}\vec{r}. \tag{5.13}$$

Hence m, as well as M, will move in an elliptical orbit with respect to C.

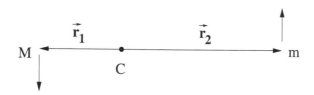

FIGURE 5.4: Two particles of mass M and m moving under their mutual gravitational field. Here C denotes their center of mass, and $\vec{r}_1$ and $\vec{r}_2$ the position vectors of M and m, respectively, with respect to C.

5.2 Application to Solar System

The result obtained in Equation 5.8 can be applied approximately to the solar system. It describes the trajectory of all the planets fairly accurately. The corrections to Equation 5.8 arise because the Sun and planets are not point particles and furthermore the solar system contains more than two objects. The extended nature of Earth results in additional gravitational effects, such as tides. Furthermore, precession of the equatorial plane of Earth arises due to torques generated by its small deviation from spherical symmetry. The effects of other planets and moons on the motion of a particular planet can be treated as a small perturbation.

The laws governing the motion of planets in the solar system were first obtained observationally by Kepler in the early seventeenth century. It was later shown by Newton that these can be derived by applying the law of gravitation, Equation 5.1, to the Sun-planet system. Kepler's laws can be stated as follows:

1. Each planet moves in an elliptical orbit with the Sun at one of its foci.

2. The line joining the Sun and the planet traces out equal areas in equal

time intervals. This law follows from the conservation of angular momentum. Let us assume that mass m moves from point A to B along an elliptical trajectory, as shown in Figure 5.5, in a small time Δt. The angle traversed in this time is $\Delta\theta$. The area swept during this time interval is the area of the triangle OAB, given by $\Delta S = r^2\Delta\theta/2$. Hence the rate of change of area is

$$\frac{dS}{dt} = \frac{1}{2}r^2\frac{d\theta}{dt}.$$ (5.14)

This is constant because $L = mr^2(d\theta/dt)$ is constant.

3. The time period T of the orbit is proportional to the cube of the semi-major axis, that is, $T^2 = Ca^3$, with the constant of proportionality C the same for all planets. From Equation 5.9 we see that this law follows approximately from the exact equations with $C = 4\pi^2/(MG)$.

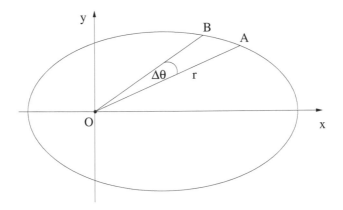

FIGURE 5.5: A particle moving in an elliptical orbit due to gravitational force of a very heavy particle located at the origin. In a small time interval Δt, it moves from A to B tracing out an area equal to $r^2\Delta\theta/2$.

5.3   Virial Theorem

Let us consider a system of N particles with masses m_i and position vectors $\vec{r}_i$ with $i = 1, 2..., N$. We define the virial of the system as

$$V = \sum_{i=1}^{N} m_i \vec{r}_i \cdot \frac{d\vec{r}_i}{dt}.$$ (5.15)

Taking the time derivative, we obtain

$$\frac{dV}{dt} = \sum_{i=1}^{N} m_i \frac{d\vec{r}_i}{dt} \cdot \frac{d\vec{r}_i}{dt} + \sum_{i=1}^{N} m_i \vec{r}_i \cdot \frac{d^2\vec{r}_i}{dt^2} = 2K + \sum_{i=1}^{N} \vec{r}_i \cdot \vec{F}_i \,, \tag{5.16}$$

where $\vec{F}_i = m_i(d^2\vec{r}_i/dt^2)$ is the total force acting on the ith particle and K is the kinetic energy of the system. Let $\vec{F}_{ij}$ denote the force exerted by the particle j on i. A particle does not exert any force on itself. Hence let us set $\vec{F}_{ii} = 0$. We can express the second term on the right-hand side as

$$\sum_{i=1}^{N} \vec{r}_i \cdot \vec{F}_i = \sum_{i=1}^{N} \sum_{j=1}^{N} \vec{r}_i \cdot \vec{F}_{ij} = \frac{1}{2} \sum_{i=1}^{N} \sum_{j=1}^{N} \left[\vec{r}_i \cdot \vec{F}_{ij} + \vec{r}_j \cdot \vec{F}_{ji} \right] \,,$$

where the two terms on the right-hand side are equal to one another. They differ only by the interchange of indices, $i \leftrightarrow j$. This does not produce any change because these are dummy indices. Using Newton's third law of motion, we have, $\vec{F}_{ij} = -\vec{F}_{ji}$. This leads to

$$\sum_{i=1}^{N} \vec{r}_i \cdot \vec{F}_i = \frac{1}{2} \sum_{i=1}^{N} \sum_{j=1}^{N} (\vec{r}_i - \vec{r}_j) \cdot \vec{F}_{ij} \,.$$

Let us now specialize to the gravitational force. In this case,

$$\vec{F}_{ij} = -\frac{Gm_i m_j}{|\vec{r}_i - \vec{r}_j|^3} (\vec{r}_i - \vec{r}_j) \tag{5.17}$$

for $i \neq j$. Hence we obtain

$$\sum_{i=1}^{N} \vec{r}_i \cdot \vec{F}_i = -\frac{1}{2} \sum_{i=1}^{N} \sum_{j=1}^{N} \frac{Gm_i m_j}{|\vec{r}_i - \vec{r}_j|} \bigg|_{i \neq j} \,. \tag{5.18}$$

The sum on the right-hand side excludes terms corresponding to $i = j$. The potential energy, U_{ij}, of a pair of particles corresponding to index i and j is given by

$$U_{ij} = -\frac{Gm_i m_j}{|\vec{r}_i - \vec{r}_j|} \,. \tag{5.19}$$

Hence the right-hand side of Equation 5.18 is the sum of potential energies of all pairs of particles. Note that each pair is counted twice. Hence the sum, including the factor of $1/2$, is simply the total potential energy of the system. Substituting in Equation 5.16, we obtain

$$\frac{dV}{dt} = 2K + U \,. \tag{5.20}$$

We next take the average of Equation 5.20 over a long time period T. The time average of a variable X, $\langle X \rangle$, is defined as

$$\langle X \rangle = \frac{1}{T} \int_0^T dt X \,. \tag{5.21}$$

The time average of Equation 5.20 leads to

$$\frac{1}{T}\left[V(T) - V(0)\right] = 2\langle K \rangle + \langle U \rangle. \tag{5.22}$$

Let us now assume that we are dealing with a periodic system or a system in equilibrium. In either case, $V(T) \approx V(0)$ and, hence, for large time T, the term on the left-hand side of this equation is approximately zero. Hence we obtain

$$2\langle K \rangle + \langle U \rangle = 0. \tag{5.23}$$

This is called the virial theorem.

An important application of this theorem arises when analyzing the evolution of large gravitationally bound systems, such as clouds of gas and dust in spiral galaxies whose collapse leads to the formation of stars. These clouds normally maintain equilibrium, in which case $2\langle K \rangle + \langle U \rangle \approx 0$. At some stage they might start collapsing due to some perturbation, such as a supernova explosion in the vicinity. The cloud starts collapsing and the virial theorem is no longer applicable. In this case, $2\langle K \rangle < -\langle U \rangle$. Eventually the heat generated due to collapse leads to an increase in $\langle K \rangle$ and again maintains equilibrium.

5.4   Tidal Forces and Roche Limit

In Section 5.1 we discussed the motion of two point particles under their mutual gravitational attraction. The situation is more complicated if we consider extended objects. In this case the gravitational force is different at different points on the object. Hence it can distort the shape of the object or in some cases can tear it apart. These force differentials are responsible for the tides observed on Earth and hence are called tidal forces.

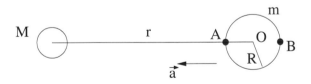

FIGURE 5.6: A uniform spherical object of mass m in the gravitational field of another object of mass M. The radius of object of mass m is R. Different segments on m, such as the two point masses located at A and B, experience tidal forces. The arrow indicates the direction of the acceleration vector, $\vec{a}$, of the center of mass of object m.

Let us consider the tidal force on a uniform sphere of mass m and radius R that is located at a distance r from a uniform sphere of mass M. Consider a point mass μ located at position A, as shown in Figure 5.6. The difference in magnitude of the gravitational field, Equation 5.3, of M between A and the center O is equal to

$$\Delta g = \frac{GM}{(r-R)^2} - \frac{GM}{r^2} \approx \frac{2GMR}{r^3} + \dots, \qquad (5.24)$$

where we have expanded the denominator $(r-R)^2$ in powers of R/r and kept only the leading order term. The direction of the vector $\Delta \vec{g}$ is toward M. Hence a particle at position A will have a tendency to get pulled away from m due to this force. Similarly, as an exercise, you may show that a particle at B will also get pulled away from m with a force of the same magnitude. The direction of $\Delta \vec{g}$ at a few representative points on m is shown in Figure 5.7. We find that at C and D, $\Delta \vec{g}$ points toward the center O. Hence points at different locations will have a tendency to be pulled apart in different directions, as shown in Figure 5.7. The magnitude of $\Delta \vec{g}$ also depends on position.

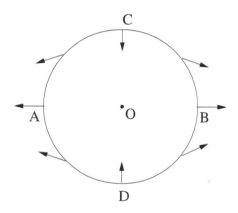

FIGURE 5.7: The direction of the tidal force at different points on a uniform spherical object of mass m moving in the gravitational field of a very heavy object M, shown in Figure 5.6. (Adapted from D. Kleppner and R. Kolenkow, *An Introduction to Mechanics.*)

The variation of $\Delta \vec{g}$ with position explains the phenomenon of tides observed on Earth. This is caused by the effect of the Moon and the Sun. For simplicity let us consider only the Sun, represented by mass M in the example discussed above. Earth is represented by mass m. Consider a small element of water, of mass μ, near the surface on an ocean. Let us first ignore the the effect of the variation of the Sun's gravitational field on Earth. In this approximation, all points on Earth accelerate toward the Sun at the same rate. Hence the pressure due to the ocean balances the gravitational pull of Earth on μ and is not affected by the Sun's gravitational field. Due to the variation in the Sun's gravitational field, however, the ocean pressure depends on position at the surface of the Earth, causing the water to rise at some point and fall at another. This leads to tides. In our discussion, for simplicity, we have ignored the contribution due to rotation of the Earth.

Let us now assume that the mass m is moving directly toward M. At some distance, the tidal forces would become so large that they might completely disrupt the object m. Its gravitational attraction may no longer be enough to hold itself together. The minimum distance m can approach M before breaking apart is called the Roche limit. Here we compute this distance by assuming that m is moving directly toward M and not rotating about its axis. We will consider a point mass μ located at A and estimate the distance r_{min} at which it begins to fly off the surface of m. For simplicity, we also assume that M is very heavy and can be taken to be at rest. All points on m undergo acceleration equal to $\vec{a}$, which is equal to the acceleration of the center of mass. By Newton's law we obtain

$$a = \frac{GM}{r^2} \, , \tag{5.25}$$

where r is the distance between the centers of the two masses. The direction of the acceleration is directly toward M as indicated in Figure 5.6. The equation of motion of μ can be written as

$$\frac{GM\mu}{(r-R)^2} + F = \mu \, a \, ,$$

where F represents the force exerted by mass m on μ. This includes the gravitational pull of m as well as the force exerted by the medium, such as normal reaction. Using Equation 5.25, we obtain

$$F = GM\mu\frac{2R}{r^3} + \dots . \tag{5.26}$$

In the limiting case when the particle is just about to fly off from the surface of m, F is equal to the gravitational attraction of m. At this point the normal reaction from the surface of m becomes equal to zero. Hence we obtain

$$\frac{Gm\mu}{R^2} = GM\mu\frac{2R}{r^3} \, ,$$

which gives

$$r^3 = 2R^3\frac{M}{m} \, .$$

This is the minimum value of r for which the mass μ will not fly off the surface of m. Let ρ and ρ_M be the densities of m and M, respectively. Let the radius of M be R_M. Hence we find

$$r_{min} = \left(\frac{2\rho_M}{\rho}\right)^{1/3} R_M \, . \tag{5.27}$$

If $r < r_{min}$, then the mass m will begin to break apart if gravitational force is the only attractive force that holds it together. This limiting value of r is called the Roche limit. One can also perform a similar calculation for an object m in circular orbit around M. The simplest case is that of an object (m) that rotates about its own axis at precisely the same angular speed as it revolves around M. In this case all points on m revolve around M with the same angular speed at a fixed distance from its center. You can analyze this system as an exercise.

In the case of objects composed of fluids, the situation is more complicated. As the object approaches another heavy object, its surface starts to distort even before it reaches the Roche limit. We will discuss some aspects of this distortion in Chapter 13. A detailed calculation of the Roche limit for such objects, taking their distortion into account, gives the result

$$r_{min} = 2.44 \left(\frac{\rho_M}{\rho} \right)^{1/3} R_M . \tag{5.28}$$

Exercises

5.1 Use Kepler's third law to determine the time period of revolution of the Earth around the Sun.

5.2 Verify the virial theorem for the case of periodic motion of two particles gravitationally bound to one another.

5.3 Verify Equation 5.24 for the difference in gravitational field at A compared to O, for mass m shown in Figure 5.6. Repeat this calculation for points B, C, and D on mass m. The points C and D are shown in Figure 5.7. Show that for B the magnitude of $\Delta \vec{g}$ is the same as at A. The magnitudes at C and D are half in comparison to that at A. The directions at all these points are shown in Figure 5.7.

5.4 Determine the Roche limit for a mass m revolving about the mass M in a circular orbit. Assume that m also rotates about its axis at the same rate as it revolves about M. In this case, all points on m move in a circular orbit around the center of M with the same angular speed. Assume a point particle placed at position A (Figure 5.6) that is bound to m only by its gravitational attraction. Find the minimum distance of m from M for which this particle will not leave the surface of m.

$$Ans: \quad r_{min} = \left(\frac{3\rho_M}{\rho} \right)^{1/3} R_M$$

Chapter 6

Stars, Stellar Spectra, and Classification

6.1 Introduction

The Universe contains a huge number of stars. Our galaxy has in excess of 200 billion stars, and there are billions of galaxies in the Universe. The density of stars is very high near the center of the Milky Way and decreases as we go outward along the galactic disk. In Figure 6.1 we show a Hubble telescope picture of the Sagittarius Star Cloud, located near the center of the Milky Way. We observe a very high density of stars with varying luminosities and colors. We also observe huge clusters of stars, called globular clusters, in the galactic halo. One example is shown in Figure 6.2. The properties of stars show wide variation in luminosity, size, mass, and spectrum. It is of course not possible to directly measure all these attributes. For most of the stars, we are only able to observe their flux density at different frequencies. Hence we know their apparent magnitudes corresponding to different filters, such as R, B, V, U, etc. Furthermore, their spectral lines can be measured very accurately. For some binary systems, the masses and radii of stars can also be directly measured. We point out that all stars, other than the Sun, appear as point-like objects. It is not possible to directly measure their angular diameters. For some stars the angular diameter measurement has been made either by interferometry or using occultations. Using flux and spectral measurements, it is possible to deduce other properties, such as their mass, radius, chemical composition, surface temperature, etc. In this and the next few chapters we study how this is accomplished.

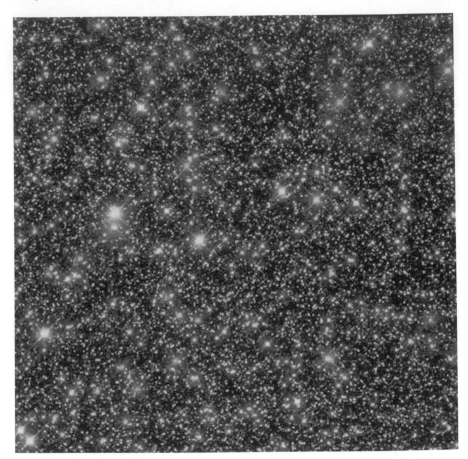

FIGURE 6.1: The rich distribution of stars in the Sagittarius Star Cloud, near the galactic center. Most of this region is not visible due to the high density of dust. This picture was taken in the direction that is relatively transparent. (Image courtesy of NASA.)

The stellar radiation approximately follows the blackbody distribution. Superimposed on this continuum radiation we observe absorption lines and in some cases emission lines. The observed solar flux density as a function of wavelength, at the top of the atmosphere, is shown in Figure 6.3. It is approximately described by a blackbody distribution corresponding to $T = 5,778$K. Superimposed on the blackbody continuum we see a large number of dips, which are absorption lines produced in stellar atmospheres. These can be seen more clearly in Figure 6.4, where we show the spectrum in a narrow range of frequencies. Historically, the physics of stars became clear only after these spectral lines were properly interpreted.

FIGURE 6.2: The globular cluster, M80. Globular clusters are found in the Milky Way halo and typically contain about 100,000 stars. (Image courtesy of NASA.)

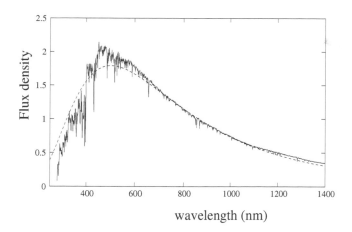

FIGURE 6.3: The solar spectrum (solid curve) as seen at the top of the atmosphere. The specific flux density, in units of $W/(m^2 \cdot nm)$, is plotted as a function of the wavelength in nanometers. The plot shows continuum radiation along with a large number of absorption lines. The smooth dashed line represents a blackbody spectrum corresponding to a temperature of 5,778K.

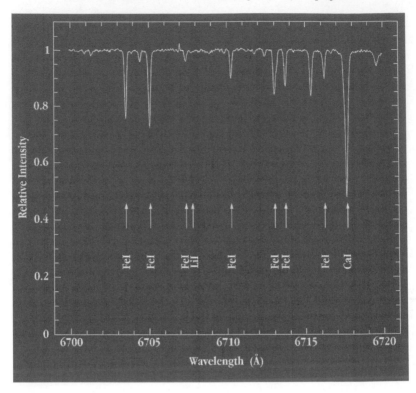

FIGURE 6.4: A sample of the solar spectra, showing absorption lines in a narrow range of frequencies. The elements that lead to these lines are also indicated. This image was taken by UVES in the Laboratory in Garching on October 19, 1998. (Image courtesy of European Southern Observatory (ESO).)

6.2   Stellar Spectra

The stellar spectral lines are produced in the atmospheres of stars. The solar spectrum was first observed by Fraunhofer in 1814. He observed a large number of absorption lines in the Sun's spectrum. At that time the quantum theory of atoms had not been developed. Hence Fraunhofer did not know how these lines originated. He later found that the different stars show different spectra. The first photograph of stellar spectra was obtained by Henry Draper. He photographed the spectral lines in star Vega in 1872. He also photographed the spectrum of many more stars. The first comprehensive survey and classification of stars, based on their spectral lines, was done at Harvard by E. C. Pickering and his two assistants, W. P. Fleming and A. J. Cannon.

The classification scheme was finalized in its present form by Cannon in 1901 and is called the Harvard classification scheme. Cannon classified the spectra of about 225,000 stars. The results were published in the form of the Henry Draper catalog between the years 1918 and 1924. The classification was done on the basis of the strength of a few representative spectral lines at visible frequencies. These include the Hydrogen Balmer lines, and lines of other elements such as He, Ca, Fe, as well as molecules such as TiO. However, at that time the astronomers were not able to properly interpret the observed spectrum. The underlying reason for the difference in spectral lines among different stars was not clear. The strength of the spectral lines could depend on many factors, such as the chemical composition and physical conditions at the stellar surface.

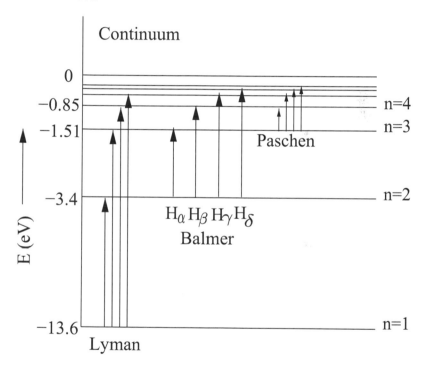

FIGURE 6.5: The energy levels and different spectral lines of the hydrogen atom. The discrete energy levels which have $E < 0$ correspond to a bound state. The continuum represents an infinite number of continuous energy levels and correspond to a free electron.

The spectra originates when an atom or an ion makes transitions among different quantum states. The different quantum states of the hydrogen atom are shown in Figure 6.5. The lowest or ground state has energy equal to −13.6 eV. An electron in the ground state can absorb an incident photon and jump to one of the excited states. Similarly, a photon in one of the excited states can absorb a photon and jump to states of higher energy. All these transitions lead to an absorption line spectrum. The energy, E, of the photon absorbed is given by $E = E_2 - E_1$, where E_1 and E_2 are the two states participating in the transition. This transition leads to an absorption line at frequency ν, given by

$$h\nu = E_2 - E_1 \,, \tag{6.1}$$

where we have used Equation 2.5. Similarly, an electron in an excited state can make a transition to a lower energy state by emitting a photon. This produces an emission line spectrum. In the case of hydrogen, the Balmer lines lie at visible frequencies and hence play a special role in astronomy.

The energy levels are discrete only for $E < 0$ which corresponds to an electron bound to a Hydrogen atom. For $E > 0$ the electron becomes free. It is no longer bound to the Hydrogen atom. In this case there is no restriction on its energy which can take any value. Hence we have an infinite number of continuous energy levels labeled as continuum in Figure 6.5.

It was only after the development of the quantum theory that it was possible to correctly interpret the origin of stellar spectra. One could then determine the physical properties of stars, such as the temperature, pressure, density, as well as the relative abundance of different elements. The stellar spectra, however, is very complicated and hence this task is rather difficult. The spectra differ for different elements and also depend on their stage of ionization. The proper framework for their interpretation is quantum statistical mechanics. The basic equations that are required are the Boltzmann distribution and the Saha ionization equation. The Boltzmann distribution forms the basis of statistical mechanics. The Saha ionization equation was developed in 1920 and allows one to compute the relative abundance of atoms in different stages of ionization. Using this equation, astrophysicists were able to establish that the Harvard classification is primarily a temperature sequence. The next important step was taken by Cecilia Payne, a doctoral student at Harvard, who performed a detailed analysis of stellar spectra in order to deduce the physical and chemical properties of stars using the Saha equation. An important outcome of her analysis was that the most common elements in stellar atmospheres are hydrogen and helium. At the time it was generally assumed that the relative abundance of elements in stars is similar to that on Earth. On Earth, helium is very rare. Moreover, elements such as oxygen, silicon, iron, calcium were found to be more abundant than hydrogen in the Earth's crust. Hence this was a very surprising discovery at the time.

We will indicate the state of ionization of the different atoms by a roman numeral. For example, HeI denotes the helium atom and HeII denotes the singly ionized helium atom. Similarly OIII denotes the doubly ionized oxygen ion. Each of these atoms or ions can be in any state of excitation.

6.3   Harvard Classification of Stellar Spectra

In the Harvard classification, the stellar spectra are classified into the following categories

$$O\ B\ A\ F\ G\ K\ M,$$

which may be remembered easily by the mnemonic, "Oh Be A Fine Guy/Girl, Kiss Me." Each of these classes is divided into subclasses by adding a numeral ranging from 0 to 9 in front of each of the letters. Hence the spectral type A is subdivided into $A0$, $A1$, ..., $A9$. The stars are classified based on the strength of the different spectral lines at visible wavelengths. For example, in the case of the hydrogen atom, only the Balmer series, which corresponds to transitions from $n = 2$ state to higher n states (see Figure 6.5), lies in the visible part of the spectrum. Hence the spectral lines of hydrogen would contribute only if the hydrogen atom has a significant probability of being in the second excited state.

The physical basis for the classification was unclear at the time it was completed. Only later, with the development of the Saha ionization equation, did it become clear that the Harvard sequence, $O\ B\ A\ F\ G\ K\ M$, corresponds to decreasing temperature. Hence the O type stars have the highest surface temperature while the M type stars are the coolest. Furthermore, within a class, the numeral sequence 0 to 9 represents decreasing temperature.

The Harvard sequence was earlier mistakenly associated with the lifetime of a star. It was based on a model where their dominant energy source was assumed to be the gravitational potential energy rather than the nuclear fusion reactions. The O type stars were believed to be in the early stage of star formation, which then cool off and later evolve into M type stars. Hence the O, B stars are referred to as the early types and the K, M stars as the late types. This model is now discredited but the late type, early type nomenclature is still used sometimes.

The O type stars are very hot and bluish in color. Their surface temperature ranges from 40,000K to 20,000K. Due to the high temperature, they show strong absorption lines from different ions such as HeII, CIII, NIII, OIII, SiIV, and SiV, as well as from the neutral helium atom HeI. A few emission lines can also be seen. The HI Balmer lines are visible but weak. An example of an O type star that can be seen by the naked eye is Meissa, which lies in the constellation Orion.

The B type stars are also very hot, with surface temperature ranging from 20,000K to 10,000K. They are also blue in color. Their spectra show neutral helium lines, which are strongest at $B2$. One also sees lines from ions such as OII, SiII, and MgII. The HI Balmer lines are relatively strong. A naked eye example of a B type star is Rigel, which is the brightest star in constellation Orion.

The A type stars are white in color and have surface temperature ranging from 10,000K to 7,500K. These have very strong hydrogen HI lines. One also sees lines of ions such as MgII and SiII. The CaII lines are also seen, but are weak. A naked eye example is Sirius, the brightest star in the night sky, in the constellation Canis Major.

The F type stars have surface temperatures in the range 7,500K to 6,000K and have a yellow-white color. The HI lines are getting weaker but still relatively strong. CaII and FeII lines start getting stronger. Neutral metal lines (FeI and CaI) also become visible. An example is Canopus in the constellation Carina. It is the second brightest star in the night sky.

The G type stars are yellow stars with surface temperatures roughly in the range 6,000K to 4,500K. Here the HI lines are weak. CaII lines continue to become stronger. FeII lines are strong. The strength of FeI and CaI lines increases from G0 to G9. The Sun is a $G2$ type star. Another example is Alpha Centauri A in the constellation Centaurus.

The K type stars are orange stars with surface temperatures in the range 4,500K to 3,600K. The HI lines are now very weak. The spectrum is dominated by metal absorption lines. CaII lines are very strong. The neutral metal lines FeI and SiI are visible. The molecular bands of TiO become visible by $K5$. An example is Aldebaran in the constellation Taurus.

The M type stars are cool red stars with surface temperatures less than 3,600K. Here the spectra are dominated by TiO bands and neutral metal (FeI) lines. The HI lines are absent. CaI lines are also very strong. An example is Betelgeuse in the constellation Orion.

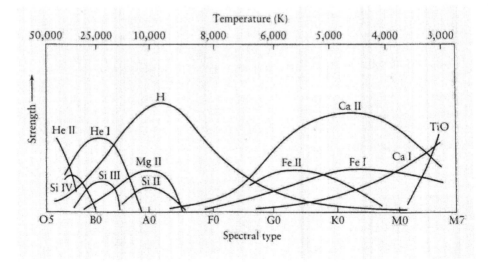

FIGURE 6.6: The relative strength of the stellar spectral lines of some of the atoms and ions. (Carroll, Bradley W.; Ostlie, Dale A., An Introduction to Modern Astrophysics, Ist Edition, 1996. Reprinted by permission of Pearson Education, Inc., Upper Saddle River, NJ.)

The strengths of some of the prominent lines as a function of the temperature are shown in Figure 6.6. The equivalent width, which is a measure of the strength of a line is shown on the y-axis. The temperature is shown along the x-axis. The spectral class corresponding to each temperature is also shown.

6.4   Saha Equation

In order to deduce the conditions at the surface of a star from the observed spectra, we need to determine how the number density of atoms in different stages of excitation and ionization depends on temperature. Once we have the theoretical framework that provides this relationship, we can determine the surface temperature from the observed spectra. The situation is complicated because the stellar atmosphere consists of many different elements in different stages of excitation or ionization. If we compare the strengths of calcium lines to those of hydrogen lines in a G type star, such as the Sun, we find that they are comparable with one another. Hence a sensible conclusion is that the relative abundance of calcium in the Sun is comparable with that of hydrogen. This was exactly what was believed up until about 1920. Then, a detailed application of Saha equation revealed that the dominant elements in the Sun and all stars are hydrogen and helium. All other elements constitute, at most, 2% by mass. In this section we explain how the Saha equation leads us to this conclusion.

The relative probability of finding atoms in different stages of excitation is given by the Boltzmann distribution. Let $P(E_a)$ be the probability that the atom is in the energy state E_a at temperature T. Let g_a be the degeneracy of this state, that is, the number of states with energy E_a. Thus the ratio:

$$\frac{P(E_b)}{P(E_a)} = \frac{g_b e^{-E_b/kT}}{g_a e^{-E_a/kT}} . \tag{6.2}$$

Here k is the Boltzmann constant. This implies that at very low temperatures, all the atoms are in their ground state. As the temperature increases, some of the atoms go to their excited states. As the temperature increases further, the atoms may ionize. During the process of ionization, an atom releases one or more electrons, which now behave as free particles. This essentially involves a transition of a bound electron to one of the states in the continuum.

The relative number of atoms in different stages of ionization is given by the Saha equation. This equation along with the Boltzmann distribution provide the basic tools to understand stellar spectra. It is derived by assuming that the ionization process behaves like a chemical reaction. Let X_I represent an atom in the Ith stage of ionization. The ionization process can be represented as

$$X_I \rightarrow X_{I+1} + e^- . \tag{6.3}$$

Let N_I be the number of atoms in the Ith stage of ionization at temperature T and χ_I their ionization energy. It is the energy required to barely remove an electron in the ground state of ionization stage, I, of an atom. The Saha equation states that

$$\frac{N_{I+1}}{N_I} = 2 \frac{Z_{I+1}}{n_e Z_I} \left(\frac{2\pi m_e kT}{h^2} \right)^{3/2} e^{-\chi_I/kT} , \qquad (6.4)$$

where Z_I is the partition function, n_e the number density of free electrons, m_e the electron mass, k the Boltzmann constant, and h is Planck's constant. The partition function is a weighted sum over the different quantum states of an atom. It can be expressed as

$$Z = \sum_{a=1}^{\infty} g_a e^{-(E_a - E_1)/kT} . \qquad (6.5)$$

We will later sketch the derivation of this equation.

We next apply the Saha equation to determine the proportion of hydrogen atoms in the first excited state to the total number of hydrogen atoms and ions in a stellar atmosphere. Let's assume that the atmosphere consists of pure hydrogen. We are interested in determining the degree of ionization in such an atmosphere. A hydrogen atom has only one bound electron. Hence there are only two possible stages of ionization, the hydrogen atom and free proton, which we denote by HI and HII, respectively. The pressure of the free electron gas, P_e, in stellar atmospheres is typically 200 dynes/cm². We can express the free electron number density, n_e, in terms of P_e by assuming that electrons behave like an ideal gas. By using the ideal gas equation of state, we obtain

$$P_e = n_e kT . \qquad (6.6)$$

We next evaluate the partition functions in stage I and II. For the hydrogen atom,

$$Z_I = g_1 + \sum_{a=2}^{\infty} g_a e^{-(E_a - E_1)/kT} .$$

At typical stellar surface temperatures, the terms corresponding to $a \geq 2$ are negligible. The degeneracy, g_1, of the ground state of hydrogen atom is 2. Hence $Z_I \approx g_1 = 2$. The partition function Z_{II} is clearly equal to unity because in this stage of ionization there are no bound electrons. Hence only one configuration is possible. The ionization energy in this case is equal to the binding energy of the hydrogen atom in the ground state, that is, 13.6 eV. Substituting all these into the Saha equation, we find

$$\frac{N_{II}}{N_I} = \frac{kT}{P_e} \left(\frac{2\pi m_e kT}{h^2} \right)^{3/2} e^{-13.6/kT} . \qquad (6.7)$$

Here N_{II} and N_I represent the number densities of hydrogen ions and neutral hydrogen atoms, respectively. In Figure 6.7 we show the ratio $N_{II}/(N_I + N_{II})$ as a function of temperature. We find that at a temperature of about 5,000K, the number of ions is negligible. Ionization increases rapidly above a temperature of 8,000K. By a temperature of about 10,000K, almost all the atoms are ionized. The important point is that ionization occurs very rapidly, over a relatively short temperature interval.

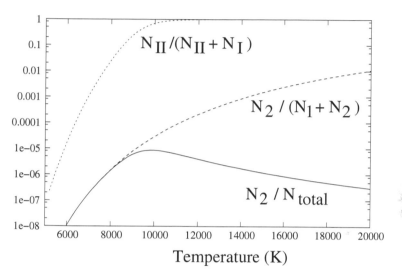

FIGURE 6.7: The ratio $N_{II}/(N_{II} + N_I)$ (small dashed line) and $N_2/(N_1 + N_2)$ (dashed line) for hydrogen. Here N_I and N_{II} represent the number of hydrogen atoms in the atomic and ionized state, respectively. The number of hydrogen atoms in the ground and first excited states are represented by N_1 and N_2, respectively. The solid line shows the ratio N_2/N_{total}, where $N_{total} = N_I + N_{II}$ is the total number of hydrogen atoms in the atomic or ionized state.

We next compute the relative number of atoms in the first excited state. Let the number densities of atoms in the ground state and first excited state be N_1 and N_2, respectively. The probability for an atom to be in an excited state is given by the Boltzmann distribution, taking into account the degeneracy of the energy level. Hence the ratio of the number of atoms in the first excited state to those in the ground state is given by

$$\frac{N_2}{N_1} = \frac{P(E_2)}{P(E_1)} = \frac{g_2 e^{-E_2/kT}}{g_1 e^{-E_1/kT}} . \tag{6.8}$$

For the hydrogen atom, we have $g_n = 2n^2$, hence $g_1 = 2$, $g_2 = 8$. The energies are $E_n = -13.6 \text{ eV}/n^2$, which gives $E_1 = -13.6 \text{ eV}$, $E_2 = -3.4 \text{ eV}$. We plot the ratio $N_2/(N_1 + N_2)$ in Figure 6.7 as a function of temperature. We find that the ratio is significantly different from zero only for temperatures greater than 10,000K. Comparing with $N_{II}/(N_I + N_{II})$, we see that by this temperature most of the atoms are ionized. Hence we find that the number of atoms in the first excited state compared to the total number of atoms (including ions) is very small. This happens because for most temperatures, the ionization probability is much higher in comparison to the probability of an atom being in an excited state.

In Figure 6.7 we also plot the number of hydrogen atoms in the first excited state in comparison to the total number of hydrogen atoms and ions. We find that the relative proportion of atoms in the first excited state is very small over the entire range of temperature. This plot can be used to interpret the stellar spectra, in particular to obtain the number density of hydrogen atoms and ions in the stellar atmosphere. Similar plots can be made for other elements. Using the observed strength of the spectral lines, these provide the densities of different elements in stellar atmospheres. The plot, Figure 6.7, holds the key to understanding why the strength of the hydrogen lines is comparable to those of many other elements even though the density of hydrogen is much larger. It shows that only a tiny fraction of hydrogen is in the right state to produce Balmer lines. Most of it is either in the ground state or ionized. Hence the effective width of the Balmer lines is very small even if the density of hydrogen is relatively large.

Let's consider the case of solar spectra. The surface temperature $T = 5,770K$. This implies that most hydrogen atoms are unionized and in their ground state. Therefore we expect a very small effective width of HI Balmer lines. Compare this with the lines of singly ionized calcium, CaII. The ionization energy of calcium is 6.11 eV. Hence the probability of ionization at solar temperature is quite high. Calculations show that practically all calcium atoms are in the ionized state, CaII. Furthermore, one also finds that most of the CaII ions are in their ground state. The transition from ground to first excited state produces the observed spectral lines in the visible region. This implies that the strength of CaII lines is much higher even if the density of calcium is much smaller than that of hydrogen.

In the entire range of temperature, the probability of producing HI Balmer lines in stellar atmospheres is very small. Hence even though the density of other atoms, such as Mg, Si, Ca, Fe, etc., is small, they produce lines of comparable or larger effective width in comparison to hydrogen.

The fact that the ionization probability is higher in comparison to the probability of an electron being in the first excited state seems counterintuitive. After all, it requires much more energy to ionize than to go to the first excited state. Hence the Boltzmann distribution would indicate a higher probability for the first excited state. However, we must also take into account the fact that ionization means transition of an electron from a bound state to any of the continuum levels. There is a continuous infinity of continuum states. In comparison, the first excited state is a discrete level with a degeneracy of 8. Hence the probability of ionization is higher because there exist many more ways in which an atom can ionize.

6.5 Derivation of the Saha Equation

We next outline the derivation of Saha equation. The derivation is slightly lengthy but involves several very interesting and important concepts. Hence I encourage the reader to go through it. We basically need to consider the ionization reaction,

$$X_I \leftrightarrow X_{I+1} + e^- , \tag{6.9}$$

in equilibrium and determine the number densities of different species, that is, the ions X_I, X_{I+1} and the electron e^-. Each of these particles is in a free state at temperature T. The number densities of particles depend on their distribution functions, which depend on the spins of the particles. Broadly, we split the particles into two categories, fermions and bosons. Fermions are particles with half integral spin in units of the Planck constant, $\hbar$, such as $\hbar/2$, $3\hbar/2$, etc. Examples of such particles are proton, neutron, and electron, all of which have spin $\hbar/2$. Bosons are particles with integral spin in units of $\hbar$, such as 0, $\hbar$, $2\hbar$, etc. Examples of such particles are the (1) photon, which has spin $\hbar$, and (2) hydrogen atom, which can have spin 0, $\hbar$, $2\hbar$, depending on the value of the orbital angular momentum of the electron and the relative spins of the electron and proton.

Consider a system of particles at temperature T. The number of particles, dN, in the energy range E to $E + dE$ is given by

$$dN = dg \, f(E), \tag{6.10}$$

where $f(E)$ is the particle distribution function and dg the statistical weight or the number of quantum states between energies E and $E + dE$. The distribution function, $f(E)$, represents the probability of finding a particle in a state at energy E. For fermions it takes the form

$$f_{FD}(E) = \frac{1}{\exp[-(\mu - E)/kT] + 1} . \tag{6.11}$$

This is called the Fermi–Dirac distribution. Here μ is the chemical potential. It is the amount by which the energy of the system changes with the addition of a particle, keeping the volume and entropy of the system constant. For a system of many species, each species has its own chemical potential. The characteristic feature of fermions is that only one such particle can occupy any particular state. Hence at zero temperature, the particles occupy all states available up to energy $E < \mu$, and all states beyond this energy remain empty. This is illustrated in Figure 6.8 where we show f_{FD} for $T = 0$ and $T > 0$.

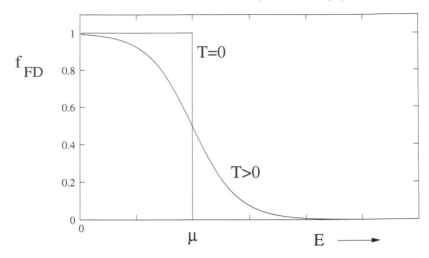

FIGURE 6.8: The Fermi–Dirac distribution f_{FD} as a function of energy E (in arbitrary units) for $T = 0$ and $T > 0$. At $T = 0$, the distribution function is equal to 1 for $E < \mu$ and equals 0 for $E > \mu$, where μ is the chemical potential.

For the case of bosons, in contrast, there is no restriction on the number of particles in any state. The distribution function takes the form

$$f_{BE}(E) = \frac{1}{\exp[-(\mu - E)/kT] - 1}, \tag{6.12}$$

with $E > \mu$. This is called the Bose–Einstein distribution. In this case, at zero temperature, all the particles occupy the lowest energy state.

For application to stars, we only need to consider the limit

$$\exp[-(\mu - E)/kT] \gg 1.$$

In this case, we find

$$dN = dg e^{\mu/kT} e^{-E/kT}. \tag{6.13}$$

For free particles in the momentum range p to $p + dp$, the number of states available is,

$$dg = \frac{g_{int}V}{h^3} 4\pi p^2 dp, \tag{6.14}$$

where V is the volume in which the particles are confined, $p = |\vec{p}|$ the magnitude of the particle momenta, and g_{int} the statistical weight of internal states of the particle. For a massive particle of spin $S\hbar$, $g_{int} = 2S + 1$. This formula does not apply to photons, which are massless. Photons have spin $\hbar$. In this case, g_{int} is equal to 2 corresponding to the two states of polarizations. We will derive the relationship, Equation 6.14 later. Using Equation 6.13 we find that the number of particles per unit volume in the momentum range p to $p + dp$ is given by

$$dn(p) = \frac{dN}{V} = g_{int}\frac{4\pi p^2 dp}{h^3}e^{\mu/kT}e^{-E/kT}. \tag{6.15}$$

Let us first compute the number density of electrons. The total energy of an electron is

$$E_e = p^2/2m_e + m_ec^2, \tag{6.16}$$

where we have included the rest mass energy, m_ec^2, because it is useful to keep track of the ionization energy of the ions, as we will see below. An electron has spin $\hbar/2$ and hence $g_{int} = 2$. Substituting this in Equation 6.15, we find

$$dn_e(p) = 2\frac{4\pi p^2 dp}{h^3}e^{\mu_e/kT}e^{-p^2/2m_ekT}e^{-m_ec^2/kT}, \tag{6.17}$$

which is just the Boltzmann distribution for a free electron gas. We now need to integrate this over p. The result is

$$n_e = 2e^{\mu_e/kT}\left(\frac{2\pi m_e kT}{h^2}\right)^{3/2}e^{-m_ec^2/kT}, \tag{6.18}$$

which can be verified by the reader. This gives the number density of electrons at temperature T.

We next consider ions in the Ith stage of ionization. Here the situation is a little complicated because the ions may be in different states of excitation and we need to consider all possibilities. Let n_{Ij} represent the number density of ions in the jth state of excitation and the Ith state of ionization. Similarly, n_{I1} represents the ions in the ground state and the Ith stage of ionization. Using the Boltzmann distribution, we find

$$\frac{n_{Ij}}{n_{I1}} = \frac{g_{Ij}}{g_{I1}}e^{-(E_{Ij}-E_{I1})/kT}. \tag{6.19}$$

Here E_{Ij} represents the energy of the ion in the jth state of excitation. The total number density of ions is

$$n_I = \sum_{j=1}^{\infty}n_{Ij} = \frac{n_{I1}}{g_{I1}}Z_I(T), \tag{6.20}$$

where Z_I is the partition function defined earlier. Here all contributions from the excited states are grouped inside the partition function. Now we only need to compute n_{I1}. We denote the mass and total energy of an ion, X_I, in the ground state by the symbols m_I and E_I, respectively. The energy is given by

$$E_I = \frac{p^2}{2m_I} + m_Ic^2, \tag{6.21}$$

where, as in the case of electrons, we have included the rest mass energy. Similarly, for the ion, X_{I+1}, we obtain

$$E_{I+1} = \frac{p^2}{2m_{I+1}} + m_{I+1}c^2 \,. \qquad (6.22)$$

Substituting this into the exponent of Equation 6.15, we find, after integrating over momentum p,

$$n_I = Z_I(T)e^{\mu_I/kT} \left(\frac{2\pi m_I kT}{h^2} \right)^{3/2} e^{-m_I c^2/kT} \,. \qquad (6.23)$$

Similarly, we obtain

$$n_{I+1} = Z_{I+1}(T)e^{\mu_{I+1}/kT} \left(\frac{2\pi m_{I+1} kT}{h^2} \right)^{3/2} e^{-m_{I+1} c^2/kT} \,. \qquad (6.24)$$

In equilibrium, we have

$$\mu_I = \mu_{I+1} + \mu_e \,. \qquad (6.25)$$

Furthermore,

$$m_I = m_{I+1} + m_e - \frac{\chi_I}{c^2} \,, \qquad (6.26)$$

where χ_I is the ionization energy, as defined earlier. We, therefore, find that the ratio

$$\frac{n_{I+1}}{n_I} = e^{-\mu_e/kT} \frac{Z_{I+1}(T)}{Z_I(T)} e^{(m_I - m_{I+1})c^2/kT} \,. \qquad (6.27)$$

In obtaining this equation, we have used the approximation, $m_I \approx m_{I+1}$, in the coefficient but not in the exponent. We can eliminate $e^{-\mu_e/kT}$ in Equation 6.27 using Equation 6.18. This finally gives us the Saha equation.

6.5.1 Number of States of a Free Particle in a Box

We next derive the relationship Equation 6.14. Here we need to determine the number of quantum mechanical states available to a particle in the momentum interval p to $p + dp$, where $p = |\vec{p}|$ is the magnitude of the momentum. The particle is assumed to be confined in a cubical box of sides L. Hence in each direction, the particle is confined to the range $-L/2$ to $L/2$. First let's consider the simpler case of a particle in just one dimension. The wave function of a free particle is equal to $\psi(x) = N \exp(ip_x x/\hbar)$, where N is a normalization factor. We impose periodic boundary conditions on the wave function,

$$\psi(L/2) = \psi(-L/2) \,. \qquad (6.28)$$

This is required so that the total probability that the particle is found inside the box is unity at all times. This leads to

$$p_x \frac{L}{2} = -p_x \frac{L}{2} + 2n_x \pi \hbar \,. \qquad (6.29)$$

Hence we find $n_x = p_x L/(2\pi\hbar)$. Therefore, the number of states in the momentum interval dp_x is given by

$$dn_x = \frac{L}{2\pi\hbar}dp_x = \frac{Ldp_x}{h}. \tag{6.30}$$

In three dimensions, the total number of states is just $dn_x dn_y dn_z$, which gives

$$dn_x dn_y dn_z = \frac{L^3}{h^3}dp_x dp_y dp_z.$$

Here $L^3 = V$ is equal to the volume of space in which the particle is confined. We are interested in the number of states in the momentum interval p to $p + dp$. Hence we should integrate over the angles in momentum space. We use $dp_x dp_y dp_z = d^3p = p^2 dp \sin\theta d\theta d\phi$. We integrate over the range $0 \le \theta \le \pi$ and $0 \le \phi < 2\pi$ to obtain

$$dg = \frac{V}{h^3}pdp \int \sin\theta d\theta d\phi = \frac{V}{h^3}4\pi p^2 dp. \tag{6.31}$$

If the particle has g_{int} degrees of freedom, then

$$dg = \frac{g_{int}V}{h^3}d^3p = \frac{g_{int}V}{h^3}4\pi p^2 dp, \tag{6.32}$$

which is the desired result.

6.6 HR Diagram

After having accumulated the temperature and absolute magnitude data for a large number of stars, astronomers tried to find out if there exists any relationship among these parameters. For example, is it possible that the luminosities of different stars may be related to some other parameter such as the temperature, mass, or size of the star? A relationship was found empirically by making a plot between the absolute visual magnitude M_V and the color index $B - V$. This plot is now known as the Hertzsprung–Russell or HR diagram, after its discoverers. We show this plot for a sample of stars in Figure 6.9.

The color index $B - V$ is directly observable and is a measure of the effective surface temperature of a star. The absolute visual magnitude is estimated from the apparent magnitude after determining the distance of the star. It provides an estimate of the luminosity L of a star. One can, therefore, plot the HR diagram in terms of the effective surface temperature and the luminosity of stars. The corresponding plot is shown in Figure 6.10. The Harvard spectroscopic classification is also labeled at the top of this plot. From left to right, it goes from the very hot O type stars to the cool M type stars.

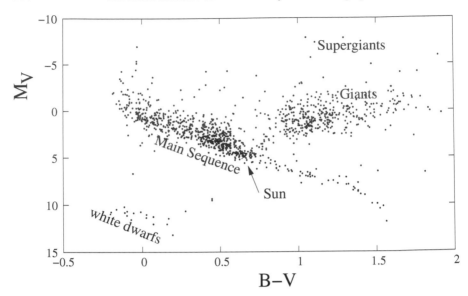

FIGURE 6.9: The HR diagram showing the absolute magnitude M_V as a function of the color index $B - V$. The Sun is located at $B - V = 0.656$ and $M_V = 4.83$. The stellar data are taken from the Hipparcos catalog.

We find that most of the stars (80% to 90%) lie roughly on a narrow diagonal band on this plot. These are called the main sequence stars. The stars in the upper left corner along this line are very hot and luminous. As we come down this line, we find that both the temperature and the luminosity decrease. Besides the main sequence, we find stars along two horizontal branches in the upper half plane of the diagram. The stars along the lower branch are called the red giants. If we compare these to the main sequence stars of comparable luminosity, we find that these stars have a lower temperature. The stars along the upper branch are called the red supergiants. These stars are considerably more luminous compared to the main sequence stars and the red giants of comparable temperature. Finally we see a branch of stars with very hot temperatures but low luminosity at the bottom of the diagram. These are called the white dwarfs.

If the star emits as a pure blackbody, then its luminosity is given by

$$L = 4\pi\sigma R^2 T^4 \,.$$

Hence, for a given temperature, the more luminous stars have larger radii. The lines of constant R are also shown in Figure 6.10. We find that these lines are roughly parallel to the main sequence band. However, there is some variation in the radius of the stars among the main sequence. The star at the extreme upper end has a radius of roughly 20 times the solar radius, whereas the one at the lower right end has a radius of about a tenth of the solar radius. The giant stars have radii lying roughly between 10 and 100 solar radii. The supergiants are larger. For example, the supergiant Betelgeuse, with effective temperature $T = 3,500$K has a radius of about 1,000 times the solar radius. This means that if it is located at the position of the Sun, then it will extend up to Jupiter. The density of the giant and supergiant stars is very small. The Sun has a density of about 1.41 g/cm^3, which is roughly the density of water. The star, Sirius, which is an $A1$ star, has a density of 0.79 g/cm^3. In comparison, the density of Betelgeuse is $10^{-8}\rho_{Sun}$, which is roughly $10^{-5}\rho_{air}$.

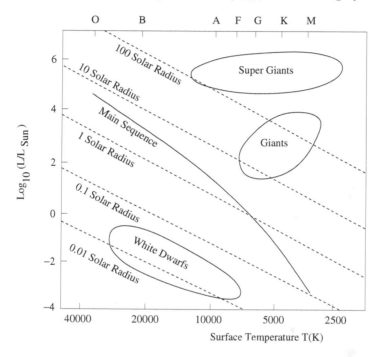

FIGURE 6.10: A representation of the HR diagram in terms of effective surface temperature and luminosity. Lines of constant radius are also shown.

Historically the development of the HR diagram happened at the beginning of the twentieth century, around the same time as the Harvard classification scheme. The first diagram was a plot of the observed quantities $B - V$ versus M_V. At that time, the physics of stars and stellar evolution was not properly understood. One did not know the composition or the surface temperature of stars. Furthermore, the source of stellar energy was not known. With the advent of the theory of stellar structure and evolution, it became clear that different branches of stars in the HR diagram, Figure 6.10, represent different stages of stellar evolution. In particular, the main sequence phase represents the earliest stage of evolution, where the energy of a star is provided by fusion of hydrogen into helium in the core of the star. After exhausting the hydrogen in its core, the star enters the giant or the supergiant phase, depending on the mass of the star. A star with a mass larger than roughly 10 solar masses evolves into a super giant phase, whereas lower mass stars settle into a giant phase. After the giant phase, a star enters its final phase of evolution, which may be a white dwarf, a neutron star, or a black hole. These objects are of very small size in comparison to stars in other phases. A neutron star or a black hole is too tiny to be represented on a HR diagram. As we discuss later, a star spends much more time on the main sequence phase in comparison to the giant phase. Hence most of the stars on the HR diagram are found to lie on the main sequence. There may be a large number of white dwarfs but most of them are too dim to be observable. The stellar structure and evolution are discussed in detail in Chapters 8 and 10.

It is clear from the HR diagram that the Harvard spectral classification, which is based only on the temperature of a star, is incomplete. For example, a red giant and a main sequence star have the same temperature but very different luminosities. One requires a two-dimensional classification that takes into account this difference. This is accomplished by the Yerkes or MKK classification scheme, which identifies several different luminosity classes, labeled by Roman numerals I, II, III, etc. The luminosity classes I, II, III, IV, and V correspond to supergiants, bright giants, normal giants, sub-giants, and main sequence stars, respectively. For example, the symbol G2I, represents a supergiant with the same surface temperature as the Sun. The Sun is classified as G2V.

6.7   Star Clusters and Associations

We observe a wide range of clusters of stars, from binary systems to globular clusters, which are huge clusters containing about 10^5 stars. Large clusters are very important for studying stellar properties and evolution because all stars within a cluster are at the same distance from us. Hence the observed difference in their properties can only be attributed to their intrinsic nature. Furthermore, as we will discuss later, stars predominantly form in clusters. Hence, in most cases, all stars within a cluster were formed at the same time. We next discuss a few prominent types of observed star clusters.

Open clusters are relatively young clusters of stars, which typically contain anywhere between ten and a few thousand stars and are predominantly found in the Milky Way disk. The stars within these clusters are relatively far apart and loosely bound to one another. Hence these clusters appear transparent in comparison to other clusters. Most of these clusters are dominated by young stars. This is because, with time, the stars within these clusters drift away and the cluster breaks up.

Globular clusters are very dense clusters of stars, seen dominantly in the galactic halo. Hence these form nearly a spherically symmetric distribution within the Milky Way. These clusters are very old. In fact, an estimate of their age gives the best lower limit on the age of the Universe.

An *association* is a group of stars that may have formed together due to the collapse of a single cloud but have drifted apart and are no longer gravitationally bound to one another. Hence stars within an association have the same age and similar proper motions. An *OB association* is a cluster of young stars that is dominated by a few, of the order of 10-100, very hot and massive O and B stars. It also contains a large number of less massive stars.

6.8 Distance and Age Determination of Clusters Using Color-Magnitude Diagram

The observed relationship between the absolute magnitudes and color index of stars allows one to determine their distances. This technique can be used effectively to determine the distance of star clusters. All the stars in a cluster are located at the same distance. We can measure their color index, $B - V$, directly by measuring the apparent magnitudes B and V. For example, the visual magnitude V of the open cluster Hyades as a function of color $B - V$ is shown in Figure 6.11.

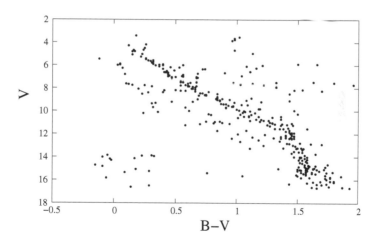

FIGURE 6.11: The color-magnitude diagram of the Hyades star cluster. Here the apparent magnitude V is plotted as a function of the color index, $B - V$.

Plotting the color and magnitude, V, of the Hyades cluster on the HR diagram, Figure 6.12, reveals useful information. We clearly identify a set of stars in this cluster that is on the main sequence. The relationship between the apparent and absolute magnitude is given by Equation 4.24. This relationship has an additional correction due to interstellar extinction, which we discuss later. We notice that because all stars in a cluster are at the same distance, the difference between m and M is a constant, equal to $-5\log(r/10 \text{ pc})$. Hence we expect that by adding a constant value to apparent magnitude of all stars in a cluster, we can get its main sequence branch to align with the corresponding branch on the HR diagram. The distance r can, therefore, be determined by matching the cluster data with the main sequence branch on the HR diagram. This procedure can be used more effectively if we use only the zero age main sequence stars (ZAMS). The relationship between absolute magnitude and color is much more precise for such stars. Hence the distance can be determined with higher accuracy. Such stars can be identified by their spectral data.

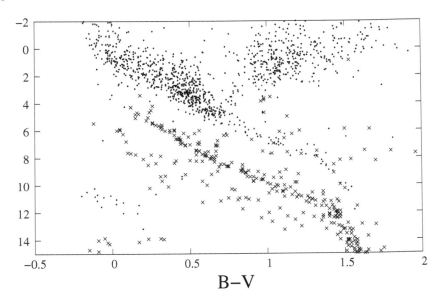

FIGURE 6.12: The color and visual magnitude V of the Hyades cluster plotted on the HR diagram, Figure 6.9. The crosses represent stars in the Hyades cluster. For these stars, the apparent magnitude V is plotted on the y-axis. The dots represent a general sample of stars whose absolute magnitude M_V is plotted on the y-axis. We can easily identify a sample of stars that lie almost parallel to the main sequence branch of the HR diagram. By adding a constant (-3) magnitude from all stars in the cluster, we can make it overlap with the main sequence. This factor contains information about the distance of the cluster. We also notice that the stars turn off from the main sequence toward the giant phase between $B - V \approx 0$. This observation can be used to estimate its age, as explained in text.

From Figure 6.12 we find that for Hyades

$$M_V \approx V - 3. \tag{6.33}$$

Hence we deduce that

$$5 \log \frac{r}{10 \text{ pc}} \approx 3. \tag{6.34}$$

We also deduce from Figure 6.12 that there are no stars in this cluster with $B - V < 0$. This suggests that at this point, the stars may have left the main sequence to approach the giant phase. This information can be used to deduce the mass of the most massive star that is still on the main sequence. The age of this star can be determined using the theory of stellar evolution. This yields an estimate of the age of the cluster. The theory of stellar evolution suggests that the luminosity L of a main sequence star, $L \propto M^{3.5}$, where M is its mass. This suggests that the lifetime t of a star on the main sequence, $t \propto 1/L^{2.5/3.5}$ (see Exercise 6.6). Using this we can express t in terms of the total time t_{Sun} the Sun will spend on the main sequence. We obtain

$$t = t_{Sun} \left(\frac{L_{Sun}}{L} \right)^{2.5/3.5}. \tag{6.35}$$

This gives an estimate of the age of the star that has just left the main sequence within a cluster. This also provides an estimate of the age of the cluster.

Exercises

6.1 Determine the wavelengths, and hence the colors, of the different Balmer lines corresponding to the transitions shown in Figure 6.5. The wavelength range corresponding to different colors can be obtained from the internet.

6.2 The star Betelgeuse has mass equal to about 8 times the solar mass and radius about 1,000 times the solar radius. Determine its density and compare with the density of the Earth's atmosphere.

6.3 Compute the ratio of the number of atoms in the first excited state, N_2, to the total number of neutral atoms, $N_1 + N_2$, at temperatures, 8,000K and 11,000K. Determine also the relative number in the ionized stage, that is, $N_{II}/(N_I + N_{II})$, at these temperatures. Finally, compute $N_2/(N_I + N_{II})$. Verify that you agree with the result plotted in Figure 6.7.

6.4 Determine the number of quantum mechanical states of a particle in two dimensions, whose momentum $p = |\vec{p}|$ lies in the range p to $p + dp$.

6.5 Using Figure 6.12, verify that you agree with the result given in Equation 6.34. Determine the distance to Hyades using this information. Compare with the best estimate of the distance to Hyades, which is equal to 47 pc.

Furthermore, estimate the age of the cluster using Equation 6.35. The lifetime of the Sun, t_{Sun}, is approximately 10^{10} years, and the absolute magnitude of a star with $B - V = 0$ is approximately 1.

6.6 Derive Equation 6.35 by using the fact that $L = E/t$, where E is the total energy emitted by a star, and the relationship $L \propto M^{3.5}$. A star generates energy by nuclear fusion. Hence the total energy emitted is proportional to M due to the mass energy relationship, $E = Mc^2$, where c is the speed of light.

Chapter 7

Radiation from Astronomical Sources

The radiation we observe from an astronomical object displays a rather complex structure. In Figure 6.3 we showed the specific flux of the radiation observed from the Sun at the top of the atmosphere as a function of the wavelength. We see a large number of sudden dips on a background continuum, which can be well fitted by a blackbody distribution with a temperature of 5,778K. These sudden drops in flux at different wavelengths represent absorption lines. Besides the blackbody spectrum, which is produced by a source in thermal equilibrium, we also find several other forms of continuum spectra.

The radiation from astronomical sources gets further distorted due to propagation through the interstellar or intergalactic medium, the interplanetary medium, the Earth's atmosphere, as well as the atmosphere of the source. The medium leads to frequency-dependent attenuation of radiation. In the visible spectrum, higher frequencies undergo larger attenuation. Hence the medium shifts the mean intensity toward smaller frequencies. The medium also produces absorption lines. In this chapter we study the production and propagation of radiation from astronomical sources.

We can broadly classify the observed spectra into the following classes:

- Continuous spectra, such as the blackbody radiation

- Discrete (very narrow) absorption and emission spectral lines

- Narrow bands composed of large number of spectral lines

We next describe each of these in detail.

7.1 Continuous Spectra

The blackbody radiation (see Figure 4.5), discussed in detail in Chapter 4, is an example of continuous emission spectra. In this case, the specific intensity shows a smooth dependence on frequency. A continuous emission spectrum is also produced by several other mechanisms besides blackbody. In some cases we may observe a continuous absorption spectrum. Let us suppose that radiation with a continuous spectral profile, such as the blackbody, passes through a medium. Some of this radiation may be absorbed and scattered by the medium, leading to a distortion in the spectral profile. If this leads to a continuous dip in the spectrum, the resulting distortion is called a continuous absorption spectrum. We discuss some standard continuous emission and absorption spectra below.

7.1.1 Synchrotron Radiation

In many astrophysical situations we have charged particles, such as electrons, moving at very high speeds in background magnetic fields. The magnetic field exerts force on these particles, causing them to accelerate. A charged particle always emits radiation when it accelerates. The intensity as well as the frequency of the emitted radiation depends on the details of the particle motion, such as its velocity and acceleration. The radiation emitted by charged particles moving at velocities close to the speed of light in background magnetic field is called synchrotron radiation. At such high velocities, the laws of classical mechanics fail and one has to use the special theory of relativity. Hence such high velocities are called relativistic. The radiation is emitted predominantly in the direction of motion of the particle. The power emitted by particles at such high speeds is very large.

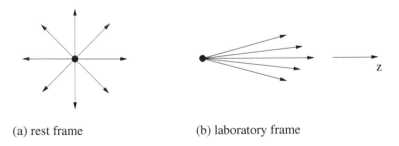

(a) rest frame   (b) laboratory frame

FIGURE 7.1: (a) In the rest frame of the electron the photons are emitted isotropically. (b) In the laboratory frame the electron moving at very high velocity along the z-axis. The photon velocities in this frame point nearly in the direction of motion of the electron.

We next explain why the radiation is emitted predominantly in the forward direction. Consider an electron moving at very high velocity in the z direction, as shown in Figure 7.1. Let us first consider emission in the rest frame of the electron. Since all directions are equivalent, the emitted radiation is isotropic in this frame (see Figure 7.1). We next transform to the laboratory frame. This can be done by adding the velocity of the electron to the velocity vectors of all the photons in the rest frame. The addition has to be done by using the special theory of relativity. It turns out that the velocities of most of the photons in this frame point nearly in the direction of motion of the electron.

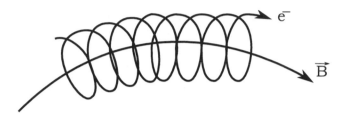

FIGURE 7.2: An electron moving in a background magnetic field undergoes a spiraling motion around the magnetic field and emits synchrotron radiation.

In a typical situation, we have a large number of electrons with different energies spiraling around the magnetic field, as shown in Figure 7.2. The electrons spiral due to the force exerted by the magnetic field. Let us assume that in some region, the magnetic field points in the z direction, as shown in Figure 7.3. Consider an electron that at some instant has velocity vector $\vec{v}$ perpendicular to the magnetic field, $\vec{B}$. The force exerted by the magnetic field on this particle is perpendicular both to $\vec{B}$ and $\vec{v}$. In the particular situation shown in Figure 7.3, the force points out of the paper. Hence the particle picks up a component of velocity that points out of the paper. The important point is that the force on the particle at every instant is perpendicular to the velocity vector. In this case the force can change the direction of motion of the particle but not its speed. The net result is that the particle starts to move in a circle.

We next consider another electron whose velocity vector makes an angle θ with respect to $\vec{B}$, as shown in Figure 7.4. In this case it is convenient to resolve the velocity vector into two components, parallel and perpendicular to $\vec{B}$. The parallel component is not affected by the magnetic force. Hence the particle moves uniformly in the z direction. However, the perpendicular component leads to a circular motion in the x-y plane. Hence the electron moves in a spiral around the z-axis.

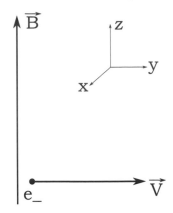

FIGURE 7.3: If an electron has velocity perpendicular to the background magnetic field, it undergoes circular motion in a plane perpendicular to the direction of the background magnetic field. Here we assume that the electron is moving in a region where the magnetic field points in the z direction. The electron will undergo circular motion in the $x - y$ plane.

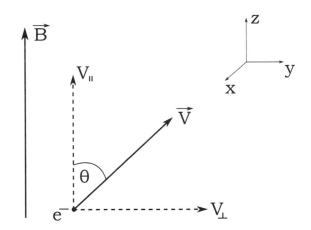

FIGURE 7.4: If an electron has velocity inclined at an angle θ with respect to the background magnetic field, it undergoes spiral motion around the background magnetic field direction. The electron's motion can be split into two parts: (1) motion with uniform speed, $v_{\parallel}$, parallel to the direction of the background magnetic field and (2) circular motion with speed $v_{\perp}$ in a plane perpendicular to the background magnetic field. These two motions combined constitute a spiral motion.

We are interested in the radiation emitted by a huge number of electrons spiraling around in background magnetic fields. The number density of electrons at a particular energy is described in terms of a distribution function

$N(E)$. Hence the number of electrons per unit volume that have energies lying in the range E to $E + \Delta E$ is equal to $N(E)\Delta E$. In many astrophysical situations, the distribution function can be expressed as

$$N(E) \propto 1/E^{\beta} , \tag{7.1}$$

where β is a constant. This is usually an approximation, valid only over a limited range of energies. Such a distribution law is called nonthermal. In contrast, a thermal distribution describes the energy distribution of a gas of particles in thermal equilibrium. In order to obtain the synchrotron power emitted by a system of electrons, we need to statistically sum the contributions over all the electrons. We skip the details here. Assuming a distribution function of the form given in Equation 7.1, we obtain the following spectral distribution of the emitted power as a function of the frequency ν,

$$f(\nu) \propto 1/\nu^{\alpha} . \tag{7.2}$$

In many cases, the exponent α is found to be approximately constant over a wide range of frequencies.

7.1.2 Bremsstrahlung

This is the radiation emitted by a free electron moving inside a plasma. A plasma is a medium consisting of free electrons and ions. A free electron moving in empty space cannot by itself emit a photon because this process violates conservation of energy and momentum. However, inside a medium, the electron can emit a photon by scattering from an ion. This is shown schematically in Figure 7.5(a). The term *Bremsstrahlung* is German for "braking radiation". The electron loses energy in the process and slows down. A free electron can have any energy, in contrast to a bound electron that can only have discrete energy levels. This implies that there is no restriction on the energy of the emitted photon. As in the case of synchrotron radiation, we need to sum over the contributions from a large number of electrons. In this case also, the power emitted shows a continuous dependence on frequency.

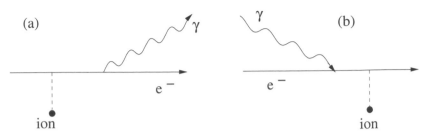

FIGURE 7.5: An illustration of (a) emission of a photon due to Bremsstrahlung and (b) absorption due to inverse Bremsstrahlung or free-free absorption. The electron scatters from an ion and absorbs or emits a photon.

In Figure 7.5(b), we show a related process called inverse Bremsstrahlung or free-free absorption. In this case, the electron absorbs radiation while scattering off from an ion. This process is applicable when we have radiation incident on some region containing plasma. The radiation gets partially absorbed and leads to distortion in its spectral distribution. It essentially leads to a reduced incident power. The amount of reduction also depends on frequency. This distortion of the spectral distribution of incident radiation is called absorption spectra.

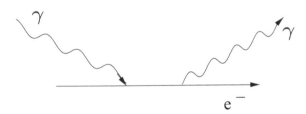

FIGURE 7.6: An illustration of the scattering of a free photon from a free electron. This process is called Compton scattering.

7.1.3   Compton Scattering

This is the radiation produced by the scattering of free photons with free electrons, as shown schematically in Figure 7.6. Here a free photon scatters on a free electron. No net photon is produced in this process. The frequency and direction of propagation of the scattered photon is different from that of the incident photon. In this case also, the scattered photons display a continuous spectrum because there is no restriction on the energy of the incident or the final photon. At low frequencies, $h\nu << m_e c^2$, this process is called Thomson scattering, discussed in more detail in the next chapter.

7.1.4 Bound-Free Transitions

A free electron in a plasma can sometimes be absorbed by an ion. The electron gets bound to the ion and occupies one of the discrete energy levels available to it inside the ion. A simple example is the capture of a free electron by a proton to form a hydrogen atom. Another example is the capture of an electron by a doubly ionized helium ion to produce a singly ionized helium ion. In this process, the electron loses energy in the form of an emitted photon. The energy of the photon is equal to the difference in the energy of free and bound electron, as shown in Figure 7.7. Because the incident electron is free, the photon produced can take a continuous range of energies. Hence this mechanism produces a continuous emission spectrum.

An atom or an ion can also absorb a photon incident upon itself. In this process, a bound electron gains energy and moves to a higher energy level, as shown in Figure 7.8. If the incident photon has sufficient energy, the electron may become free. The frequency of the photon must satisfy the relationship $h\nu > \chi$, where χ is the ionization energy of the atom or the ion. This process is called photo-ionization or bound-free transition. It produces a continuous absorption spectrum.

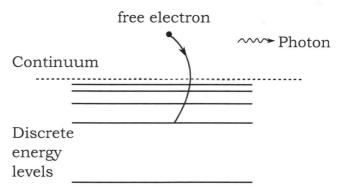

FIGURE 7.7: Continuum emission due to an electronic transition from a free state to a bound state.

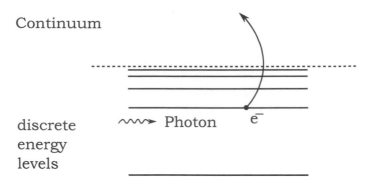

FIGURE 7.8: Continuum absorption due to an electronic transition from a bound state to a free state.

7.2 Absorption and Emission Line Spectrum

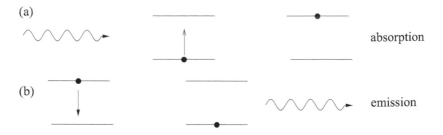

FIGURE 7.9: (a) The absorption line spectrum is generated when an atom absorbs incident continuum radiation and makes a transition to a higher excited state. (b) The emission line spectrum is produced when an atom makes a transition to a lower energy state.

We next discuss the formation and properties of discrete line spectra. In this case, the spectral profile shows sharp dips at some definite frequencies, as seen, for example, in the solar spectrum, Figure 6.3. The dips represent absorption lines. Most stars predominantly show an absorption line spectrum. These spectral lines are generated by atomic transitions. An absorption line is generated when an atom absorbs radiation from an incident continuous spectrum and makes a transition to a higher energy level as shown in Figure 7.9. An emission line spectrum is also visible in some cases. In this case, one observes a sharp peak instead of a dip. The emission line is generated when the atom in a higher excited state makes a transition to a lower energy state.

Each atom has a distinctive spectrum that depends on its energy levels. The frequency of the radiation emitted or absorbed is related to the level spacing by Equation 6.1. The spectrum also depends on physical conditions such as the temperature and pressure of the medium. The temperature, in particular, very strongly influences the spectrum because it determines which energy levels are occupied and also the stage of ionization of the atom. The resulting spectra of a star, which is composed of many different elements, in different stages of ionization, is very complex.

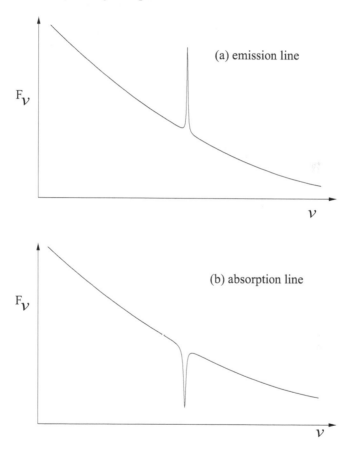

FIGURE 7.10: The flux density F_ν as a function of frequency ν for (a) an emission line and (b) an absorption line along with the background continuum radiation.

An emission and an absorption line along with the background continuum radiation is schematically shown in Figure 7.10. Here we have focused on just one line in the background continuous spectrum. The characteristics of the line are better represented in Figure 7.11 where we plot the observed flux density $F(\nu)$ divided by the background continuum radiation flux $B(\nu)$. In making this plot, we need to know the form of the background continuum. In some cases, we may know this function theoretically. Alternatively, we can obtain it by making an empirical fit to the data. As we see from Figure 7.11, the line has a finite width, which is often much smaller than the central frequency of the line. Later we will describe the physical processes that lead to a finite width of the spectral lines. We first mathematically characterize the line profile.

The normalized flux density, that is, the flux density, $F(\nu)$, divided by the background continuum, $B(\nu)$, can be expressed as

$$\frac{F(\nu)}{B(\nu)} = 1 \pm f(\nu), \qquad (7.3)$$

where the positive sign refers to the emission line and the negative sign to an absorption line. The function $f(\nu)$ characterizes the actual shape of the line. It is significantly different from zero only in a narrow range of frequencies. In Figure 7.11 we plot $F(\nu)/B(\nu)$ for an absorption line, corresponding to a negative sign in Equation 7.3. Notice that $F(\nu)/B(\nu)$ approaches unity as we move away from the spectral line. In Figure 7.12 we plot the line profile, $f(\nu)$, as defined in Equation 7.3.

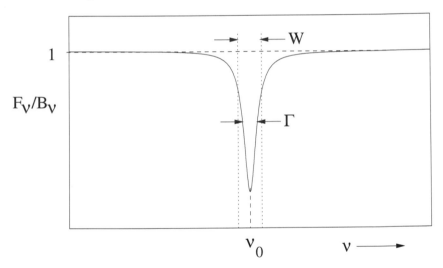

FIGURE 7.11: The flux density divided by the continuum flux F_ν/B_ν for an absorption line. The peak frequency of the line is ν_0 and its width Γ. The equivalent width W is also shown.

The shape of a spectral line is well characterized in terms of two parameters:

(1) The equivalent width W: This provides information about the overall strength of the line, as shown in Figure 7.11. It is equal to the area under the spectral line, that is, the area of the shaded region in Figure 7.12.

This area has dimensions of frequency because $f(\nu)$ is dimensionless. Hence we define an equivalent width W that is numerically equal to this area, as shown in Figure 7.11.

(2) The width at half maximum, Γ: This is the width of a spectral line at half its peak value. This parameter gives us an estimate of how narrow, or broad, the line is.

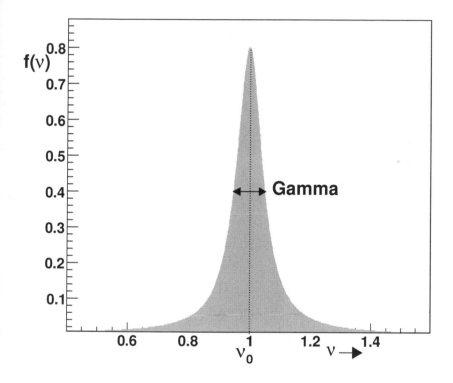

FIGURE 7.12: The spectral line profile $f(\nu)$ as a function of the frequency ν. The width of the line at half maximum, Γ (Gamma), is also shown.

Theoretically, the shape of a spectral line, ignoring the continuum, is well represented in most cases by

$$f(\nu) = WL(\nu), \tag{7.4}$$

where the equivalent width W parameterizes the strength of the line and $L(\nu)$ is the Lorentzian function, defined as

$$L(\nu) = \frac{1}{\pi} \frac{(\Gamma/2)}{(\nu - \nu_0)^2 + (\Gamma/2)^2}. \tag{7.5}$$

Here the parameter Γ characterizes the width of the line and ν_0 its peak position. Mathematically, the equivalent width is defined as

$$W = \int_0^\infty f(\nu)d\nu \,. \tag{7.6}$$

The integral of the Lorentzian function, $L(\nu)$, over the range $-\infty < \nu < \infty$, is equal to one. In our case, $\nu \geq 0$. In the limit of very small width, the integral in the range $0 \leq \nu < \infty$ is also unity, to a very good approximation. Hence if we substitute $f(\nu) = WL(\nu)$ on the right-hand side of Equation 7.6, the integral is approximately equal to W.

A sample spectral line profile $f(\nu)$ is shown in Figure 7.12. In this figure we have chosen the parameters $\nu_0 = 1$, $W = \pi/25$, and $\Gamma = 0.1$, in arbitrary or unspecified units. The flux density $f(\nu)$ at its peak position $\nu = \nu_0$ is equal to $2W/(\pi\Gamma)$. This is equal to 0.8 in Figure 7.12. The intensity reaches half its peak value at frequencies $(\nu_0 \pm \Gamma/2)$. Hence Γ is the width of the line at half its maximum value, as shown in Figure 7.12. We point out that the width Γ is typically very small in comparison to the peak frequency ν_0 of the line. Indeed, the formula for the line profile, Equation 7.4, is applicable only in the limit $\Gamma << \nu_0$.

For an absorption line, the minimum value of the spectral profile $F(\nu)$ can be zero if the medium is such that the entire continuum radiation is absorbed at the central frequency ν_0. If the value of $F(\nu)$ at ν_0 is greater than zero, then we call the medium optically thin, else it is optically thick. In the latter case, the medium is highly opaque and the radiation is completely attenuated near the central frequency.

7.2.1   Radial Velocity Due to Doppler Effect

As discussed in Chapter 3, the observed wavelength of radiation becomes smaller (blue shifted) if the source is moving toward the observer. It becomes larger (red shifted) if the source is moving away. This effect can be used to measure the velocity component of the source along the line of sight. Let us assume that the observed wavelength of a spectral line received from a distant source is λ. We do not have any direct means to estimate the emitted wavelength, λ_0, and hence we cannot determine the Doppler shift, $\Delta\lambda = \lambda - \lambda_0$ directly. However, we can deduce λ_0 by matching it with a known spectral line of some element. We can then estimate $\Delta\lambda$ from the observed λ and use Equation 3.9 to extract v_r. This measurement is best done by observing several different spectral lines. Let us assume that we observe different lines at wavelengths $\lambda_1, \lambda_2, \lambda_3, ...$, etc. We match each of these with known spectral lines, $\lambda_{01}, \lambda_{02}, \lambda_{03}, ...$, respectively. We can then deduce the Doppler shifts $\Delta\lambda_1, \Delta\lambda_2, \Delta\lambda_3, ...$, etc. and extract the corresponding velocities by using Equation 3.9. For consistency, the values of v_r extracted from different spectral lines should match one another within experimental errors. If this is found to be true, we can attribute the observed shift to Doppler effect and make a reliable measurement of v_r.

7.2.2 Causes of Finite Width of Spectral Lines

We next discuss different mechanisms that lead to a finite width of the spectral lines. A transition between two discrete energy levels will produce a spectral line at a particular frequency, which corresponds precisely to the energy difference between the two levels. However, the observed spectral lines display a finite width. This broadening of spectral lines arises due to several reasons.

1. Natural Broadening: The energy states of any atom have small widths. The width is related to the lifetime of the state by the Heisenberg uncertainty relationship

$$\Delta E \approx \hbar/\Delta t. \tag{7.7}$$

   If a state has absolutely well-defined energy, then $\Delta E = 0$ and $\Delta t = \infty$. Such a state necessarily has infinite lifetime. However, most states have a finite lifetime that leads to a small uncertainty in their energy.

2. Doppler Broadening: The atoms in stellar atmospheres are always in random motion. Due to the velocity of the atoms, the lines produced get Doppler shifted. The observed line profile arises due to a large number of atoms in random motion. Because each atom produces a line at a slightly different position compared to other atoms, the observed line profile appears broadened.

3. Pressure Broadening: The energy levels of atoms and ions get slightly shifted due to interaction with other particles in the medium. This leads to a shift in the spectral lines. The shift depends on the medium properties and changes randomly with position. Hence this also leads to broadening of the observed spectral lines.

7.3 Molecular Band Spectra

The atoms produce isolated spectral lines due to electronic transitions between different energy levels. A molecule has additional degrees of freedom as compared to an atom. The different atoms inside a molecule can vibrate about their mean positions, as shown schematically in Figure 7.13(a). A molecule can also undergo rotation, as shown in Figure 7.13(b). These lead to additional states besides the different electronic states. The transitions among these states, including the electronic transitions, produce narrow bands consisting of a large number of lines. These are called the molecular band spectra.

(a)   vibrations   (b)   rotations

FIGURE 7.13: An illustration of (a) vibrational and (b) rotational modes of a molecule.

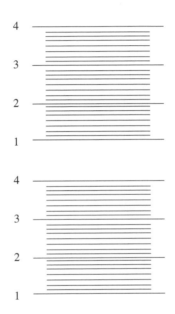

FIGURE 7.14: An illustration of the energy levels of a molecule. We show two electronic states, each of which further split into several vibrational and rotational states. Here we show four vibrational states for each electronic level. This are labeled as 1, 2, 3, and 4. The closely spaced narrower lines indicate the rotational states.

An illustration of the energy levels of a molecule is shown in Figure 7.14. Here we show only two electronic states. The molecule can undergo vibrations and rotations in each electronic state. Hence each such state has additional closely spaced states corresponding to these modes. These are indicated by the two sets of states in Figure 7.14. Typically the level splittings between vibrational states is much smaller than those between different electronic states. Transitions between different vibrational modes typically produce bands in infrared frequencies. The rotational states have even smaller energy splittings. The corresponding transitions produce bands in microwave frequencies. One can also have transitions from the vibrational or rotational states of one electronic state to those of another electronic state. These can produce bands in infrared, visible, or ultraviolet frequencies.

7.4 Extinction

The light or radiation that reaches us from an astronomical source undergoes considerable attenuation during propagation due to scattering and absorption. Due to scattering, the incident radiation is scattered by the medium in all directions. This effectively reduces the intensity of the beam reaching the observer. In the case of absorption, radiation is absorbed by the medium and then re-emitted in different directions and at different frequencies, again effectively reducing the intensity of the beam. The combined loss is called extinction.

FIGURE 7.15: A schematic illustration of the incident beam and target particles, denoted by small spherical objects. An incident particle mostly encounters empty space and occasionally scatters from a target particle.

We next mathematically describe the attenuation of radiation due to extinction. In most situations considered in astrophysics, the medium has very low density. The intensity of the incident beam is also normally very small. It is convenient to use the photon, that is, particle, interpretation of light. We imagine light as a beam of photons with flux N. The flux N is defined as the number of photons crossing a unit cross-sectional area per unit time. The flux is proportional to the intensity of light,

$$N \propto I.$$

The photons interact with the particles in the medium, which we refer to as the target particles, and are scattered and absorbed. For simplicity, let's assume that the target particles are at rest. Let their number density, that is, the number of particles per unit volume, be denoted by n. We are interested in the change in flux ΔN as the beam travels a small distance Δs in the medium.

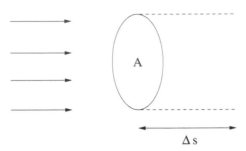

FIGURE 7.16: The incident beam of cross-sectional area A traversing a distance Δs in medium.

We can visualize extinction by considering a parallel beam of photons propagating in a medium, as shown in Figure 7.15. We have assumed that the medium is rare and the photon beam flux is relatively low. Hence as a photon propagates through the medium, it mostly encounters empty space, and the probability of getting scattered is small. Some of the photons get scattered or absorbed after colliding with the particles in the medium. Each beam particle interacts with the target particle independent of other beam or target particles. Hence the change in flux is proportional to the incident flux, the target density, and the distance traveled Δs. This can be expressed mathematically by the equation

$$\Delta N = -\sigma N n (\Delta s) , \tag{7.8}$$

where the proportionality constant σ is called the cross section and has dimensions of area. The dimensions of σ can be easily determined by noticing that ΔN has the same dimensions as N, and $n\Delta s$ has dimensions of inverse area. The negative sign in this equation means that the flux is decreasing. Hence the change in flux, ΔN, is negative. We physically interpret σ as the cross-sectional area over which the incident particle scatters or gets absorbed by a target. We can visualize this by imagining target particles as small spheres of cross-sectional area σ as shown in Figure 7.15. If the incident particle strikes within this area, then it gets scattered or absorbed, or else it propagates through unaffected.

We now derive the equation for ΔN, Equation 7.8, more directly. Let a photon beam of cross-sectional area A traverse a distance Δs in the target, as shown in Figure 7.16. The total number of target particles in volume $A\Delta s$ is equal to $nA\Delta s$ (number density times the volume). Each target particle presents an effective cross-sectional area σ over which the incident particle undergoes scattering and/or absorption. Multiplying this by the number of target particles, we find the total area over which an incident particle can interact with a target particle, within a distance Δs, is equal to $\sigma n A \Delta s$. Because there are N incident particles per unit area per unit time, the total number of particles scattered and/or absorbed per unit time in area A is $\sigma N n A \Delta s$. Hence the rate per unit area at which the particles are being removed from the incident beam is $\sigma N n \Delta s$, which gives us the final result, Equation 7.8. This relation is valid as long as the scattering and absorption of a photon with a target particle occurs independently of other target particles or photons. This assumption may be violated in the case of high target density and/or high incident flux. In this case, Equation 7.8 would be more complicated and the linear dependence on N and/or n may no longer be valid.

As a simple example, consider a particle of negligible size, such as a marble, incident on a spherical object, such as a billiard ball, of radius r, as shown in Figure 7.17. Let the impact parameter be equal to b. The impact parameter is defined as follows: Draw a perpendicular from the center of the target particle to the line along the direction of velocity of the incident particle, as shown in Figure 7.17. The length of this perpendicular line is the impact parameter. It is clear that the marble will undergo scattering if $b < r$. Otherwise it will not scatter. Therefore, the cross-sectional area for scattering, σ, is equal to πr^2.

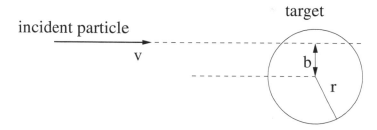

FIGURE 7.17: The incident particle, of negligible dimensions, approaches the target with velocity v and impact parameter b. The target is assumed to be a sphere of radius r.

If we consider the scattering of a photon on an electron, then the cross section does not refer to the physical size of these particles. To the best of our knowledge, both the electron and photon are point particles. However, the photon scatters on an electron due to their electromagnetic interaction through the process called Compton scattering. In this case, the cross section is equal to the effective area over which scattering occurs. Some of the photons may also be absorbed by the target particles. The probability for this process is mathematically described by absorption cross section, which is equal to the cross-sectional area over which the incident particles are absorbed by the target particles.

The cross-sectional area over which the incident photon scatters or is absorbed by a target is usually very small. A convenient unit of cross section is the barn,

$$1 \text{ barn} = 10^{-24} \text{ cm}^2. \tag{7.9}$$

In the above derivation, we have used the photon interpretation of the electromagnetic field because it is convenient. Going back to the wave interpretation, we note that the intensity of the wave I is directly proportional to N. Hence in terms of intensity, we obtain

$$\Delta I = -n\sigma I \Delta s, \tag{7.10}$$

where I represents the intensity of light and ΔI the change in intensity after propagation through a distance Δs. The equation can also be expressed in terms of the specific intensity, I_λ, by replacing I and ΔI by I_λ and ΔI_λ, respectively. It is convenient to define the absorption coefficient or opacity κ_λ, such that

$$\rho \kappa_\lambda = n \sigma_\lambda, \tag{7.11}$$

where σ_λ is the cross section at wavelength λ and ρ is the density of the medium. In terms of κ_λ, the extinction equation for specific intensity can be expressed as

$$\Delta I_\lambda = -\rho \kappa_\lambda I_\lambda \Delta s. \tag{7.12}$$

The basic difference here from Equation 7.10 is that the density ρ appears in this equation instead of the number density n. Furthermore, this equation gives us the change in intensity ΔI_λ at a particular wavelength λ. The opacity κ_λ in Equation 7.12 is a measure of how strongly the incident beam intensity attenuates in a medium. Finally we define the optical thickness τ_λ such that

$$\Delta \tau_\lambda = \rho \kappa_\lambda \Delta s. \tag{7.13}$$

In the limit of infinitesimally small distance, $\Delta s \to ds$, Equation 7.12 can be written as, $dI_\lambda = -\rho \kappa_\lambda ds$. Hence we obtain the differential equation

$$\frac{dI_\lambda}{ds} = -\rho \kappa_\lambda I_\lambda. \tag{7.14}$$

The solution to this equation can be expressed as

$$I_\lambda(s) = I_\lambda(0) \exp\left(-\int_0^s \kappa_\lambda \rho ds\right) = I_\lambda(0) \exp(-\tau_\lambda), \tag{7.15}$$

where τ_λ is the total optical thickness of the medium of length s. This describes the decrease in intensity of radiation due to absorption and scattering during propagation in a medium. As discussed in Chapter 4, in astronomy we are interested in the intensity observed through a particular filter. Hence in astronomical applications, we need to add up the contributions to Equation 7.15 due to all wavelengths corresponding to the filter.

We point out that besides being attenuated, the radiation intensity may also increase during propagation in a medium. This is due to emission from the medium. Furthermore, radiation emitted by stars, other than the source under observation, may be scattered in the direction of the observer as shown in Figure 7.18. Taking this increase into account, we can express the net change in specific intensity as

$$\frac{dI_\lambda}{ds} = -\rho\kappa_\lambda I_\lambda + j_\lambda, \tag{7.16}$$

where j_λ is called the emission coefficient of the medium. The second term on the right-hand side leads to an increase in intensity due to emission and scattering in the direction of propagation. Equation 7.16 is the general equation of radiative transfer in a medium.

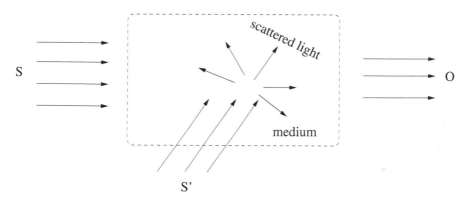

FIGURE 7.18: The radiation emitted by the source S is attenuated as it propagates in a medium toward the observer O. However, the radiation intensity also is enhanced due to scattering of radiation from other sources such as S' in the direction of O.

7.4.1 Extinction Coefficient

We next obtain a formula for the distance modulus by taking the effect of extinction into account. Consider a star of radius R. Let $F_{\lambda 0}$ be the flux density at the surface of a star at wavelength λ and $F_\lambda(r)$ the corresponding flux density at a distance $r > R$ from the center of the star. Because the flux density decreases as $1/r^2$ with distance in free space, we find that

$$F_\lambda(r) = F_{\lambda 0}\frac{R^2}{r^2}e^{-\tau_\lambda}, \tag{7.17}$$

This equation must be applied to the flux observed through a particular filter. Using the formulae for apparent and absolute magnitudes, we find that the distance modulus at wavelength λ,

$$m - M = -2.5\log_{10}\frac{F_\lambda(r)}{F_\lambda(10)} = -2.5\log_{10}\left(\frac{10\text{ pc}}{r}\right)^2 - 2.5\log e^{-\tau_\lambda}.$$

This gives

$$m - M = 5\log\frac{r}{10 \text{ pc}} + A,\qquad(7.18)$$

where the extinction coefficient,

$$A = (2.5\log_{10} e)\tau_\lambda,\qquad(7.19)$$

gives the change in the apparent magnitude due to extinction. Here τ_λ is the optical thickness of the medium between the surface of the star and the observer. We can represent it in the form of an integral as

$$\tau_\lambda = \int_R^r dr'\kappa_\lambda\rho.$$

In the formula for distance modulus, the flux density $F_\lambda(10)$ is computed by assuming that the medium is free space. Hence in this case, we set $\tau_\lambda = 0$ in using Equation 7.17 to compute $F_\lambda(10)$.

We can now express the result given in Equation 7.18 for different filters used in astronomy. These are obtained by summing over the contributions from different wavelengths, corresponding to a particular filter. For the filters V and B, for example, we obtain

$$V = M_V + 5\log\frac{r}{10 \text{ pc}} + A_V,\qquad(7.20)$$

$$B = M_B + 5\log\frac{r}{10 \text{ pc}} + A_B,\qquad(7.21)$$

respectively, where A_V and A_B are the corresponding extinction coefficients. The visual extinction A_V of the Milky Way is approximately 1.8 magnitude for a propagation distance of 1 Kpc.

7.4.2   Color Excess

Besides extinction, the interstellar medium also causes the reddening of light. This is because blue light is scattered and absorbed more than red light. In general, the attenuation is larger at smaller wavelengths, leading to a shift in the spectrum of stars toward higher wavelengths. Due to this frequency dependence, the color index $B - V$ increases. By subtracting Equation 7.20 from Equation 7.21, we obtain

$$B - V = M_B - M_V + E_{B-V}.\qquad(7.22)$$

Here the difference $M_B - M_V$ is a measure of the intrinsic or true color of the star, and the color excess

$$E_{B-V} = A_B - A_V\qquad(7.23)$$

arises due to extinction in the interstellar medium. It has been observed that, on average, for stars, the ratio

$$R = \frac{A_V}{A_B - A_V} \approx 3.0 \,. \qquad (7.24)$$

It is found to be almost the same for all stars. This gives a measure of the reddening caused by the interstellar medium.

In Figure 7.19 we show a plot of the true colors of the main sequence stars. The color index $U - B$ is plotted as a function of $B - V$ after subtracting out the color excess. The corresponding curve for a true blackbody is also shown. We find that the color-color plot shows some deviation from blackbody behavior. This is partially attributed to extinction caused by the stellar atmospheres, which also leads to reddening. In Figure 7.19 the point S represents the observed color indices of a particular star. This point does not lie on the solid curve due to the effect of the interstellar medium. The arrow indicates the approximate position of this star on the solid line after correcting for the effect of interstellar extinction.

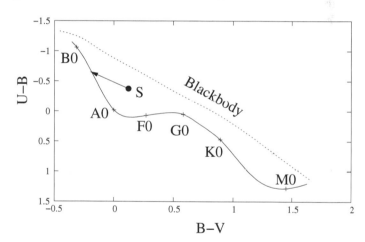

FIGURE 7.19: The color index $U - V$ as a function of $B - V$ (solid curve) for the main sequence stars. The plot shows the true color of stars after subtracting out the effect of interstellar medium. The Harvard classification of stars is also indicated. The (dashed line) shows the corresponding result for a true blackbody. The observed color indices of a star show reddening in comparison to the solid curve, as indicated by the point S. One has to apply a correction due to color excess in order to deduce the true color of the star. (Data from C. W. Allen, *Astrophysical Quantities* and H. Arp, *Astrophysical Journal* 133, 874 (1961).)

Exercises

7.1 Determine the cross section for scattering of a billiard ball from another billiard ball. Both have a radius equal to R.

7.2 Assume that a galaxy is moving away from us at a speed of 300 Km/s. Determine the observed Doppler shift of a spectral line with frequency $\nu_0 = 5 \times 10^{14}$ Hz produced by this galaxy.

7.3 Consider an atomic state that has a lifetime of 1 second. Determine the width, ΔE, of this state. A transition from this state produces a spectral line at visible frequency $\nu_0 = 5 \times 10^{14}$ Hz. Determine the width of this line caused by natural broadening. Compare this with ν_0.

7.4 Show that the solution to the differential equation Equation 7.14 is given by Equation 7.15.

7.5 The visual extinction coefficient A_V for a star at 1 Kpc is approximately 1.8 magnitude. Determine its blue extinction coefficient A_B.

7.6 The absolute magnitude M_V of a star is known to be -3.0. Its apparent visual magnitude V is observed to be 12.

   (a) Determine its distance by ignoring A_V.

   (b) If we include A_V, the distance cannot be computed directly. We can obtain a rough estimate as follows: Compute A_V for $r = 1, 2, 3, 4$ Kpc. Insert these values in Equation 7.20 and compute the distances. The value that is closest to the input distance provides a rough approximation of the true distance. This procedure can now be refined by computing A_V values over intervals of 100 pc and inserting in Equation 7.20. Notice that (a) provides a very poor approximation of the distance.

Chapter 8

Stellar Structure

The internal structure of stars cannot be probed by direct observations. We observe only the radiation emitted by the surface, which allows us to deduce the surface properties. The internal structure must be deduced indirectly by theoretical modeling. Stars are essentially gaseous structures held together by gravitational force. The outward force exerted by the gas pressure balances the inward gravitational pull to maintain equilibrium. The outward pressure is maintained by the heat generated by nuclear reactions in the core of a star.

To a good approximation, a star maintains mechanical and thermal equilibrium for most of its lifetime. The energy generated in its core does not accumulate inside the star. It slowly finds its way to the surface, where it is released in the form of blackbody radiation. Depending on the size of the star, the emission from the surface peaks at infrared, visible, or ultraviolet frequencies. The internal as well as surface properties of a star evolve very slowly. The evolution is not noticeable within the span of a human lifetime. The fact that a star evolves and emits radiation implies that it cannot be in strict mechanical and thermal equilibrium. However, the deviations are expected to be small.

After taking birth, a star goes through several different stages of evolution. It spends a maximum amount of time in the main sequence phase of its life cycle. This is the phase during which its energy is generated by fusion of hydrogen nuclei into helium. A star does undergo rapid changes when it exhausts its nuclear fuel. For example, the main sequence phase of a star terminates when the hydrogen in the core is completely converted to helium. The temperature in the core is not sufficiently high at this stage to start Helium fusion. In that case, the internal pressure is not sufficient to balance the gravitational attraction and the star tends to collapse. This initiates a sequence of events which happen rather quickly till the temperature in the core becomes sufficiently high to start nuclear fusion of the heavier elements present in the core. The physical characteristics of the star change dramatically during this process. In some cases, when it exhausts a heavier nuclear fuel such as Helium, Carbon or Oxygen, the entire star may be blown apart in a supernova explosion. The luminosity of the star rises by many orders of magnitude within a very short time scale of a few days. The only remnant left behind may be a tiny compact object called neutron star.

We next obtain the basic equations that determine the structure of stars. We shall mostly be interested in stars under equilibrium conditions. We shall assume spherical symmetry which is a very good assumption for stars. Hence the physical and chemical properties of stars depend only on the distance, r, from the center and not on the angular coordinates θ and ϕ. The basic variables describing the stellar structure are the pressure, $P(r)$, temperature, $T(r)$, density, $\rho(r)$, the energy produced within a radius r, $L(r)$ and the relative abundance of different elements. It turns out to be useful to define another variable, $M(r)$, which is equal to the mass contained within a sphere of radius r. We need equations which describe the rate of change of these variables with r.

8.1   Pressure Gradient

The basic forces that govern the dynamics of a star are

1. Gravitational attraction, which tends to collapse the star

2. The outward pressure exerted by the stellar medium

A star is in mechanical equilibrium if these two forces balance one another; otherwise the star collapses or expands. Consider an infinitesimal test element of mass Δm at a distance r from the center. Let its cross-sectional area be A and its thickness Δr, as shown in Figure 8.1. We also assume that as the star expands or contracts, the test mass does not mix with the remaining stellar medium. Hence it can be treated as an isolated system and we can apply Newton's law, $F = ma$, to this element. The mass has only radial acceleration whose magnitude is equal to $a = d^2r/dt^2$. Under equilibrium, this acceleration is equal to zero. The gravitational force acting on it is F_G. The magnitude of the force due to pressure, $P(r)$, is given by, $F_P(r) = P(r)A$. The force due to pressure acting radially on the surface at r is equal to $F_P(r)$ (see Figure 8.2). The corresponding force at $r + \Delta r$ is $-F_P(r + \Delta r)$. The negative sign in this force arises it acts radially inward. The equation of motion can now be written as

$$(\Delta m)a = F_G - F_P(r + \Delta r) + F_P(r). \qquad (8.1)$$

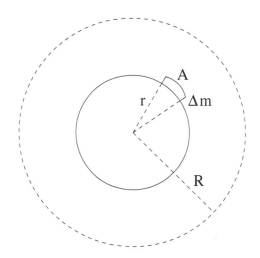

FIGURE 8.1: An infinitesimal mass element, Δm at distance r from the center of a star of radius R. The element has surface area A and thickness Δr.

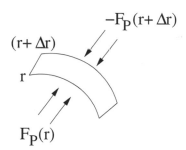

FIGURE 8.2: The force due to pressure acting on a mass element Δm of thickness Δr shown in Figure 8.1. The area of both the surfaces, located at r and $r + \Delta r$, is equal to A.

The force due to pressure is given by,

$$-F_P(r + \Delta r) + F_P(r) = -AP(r + \Delta r) + AP(r) = -A\Delta P(r), \qquad (8.2)$$

where $\Delta P = P(r + \Delta r) - P(r)$ is the difference in pressure between the two surfaces. The gravitational force can be expressed as

$$F_G = -\frac{GM(r)\Delta m}{r^2}, \tag{8.3}$$

where $M(r)$ is the total mass contained within a sphere of radius r, as shown in Figure 8.3. Therefore we can express the equation of motion as

$$(\Delta m)a = -\frac{GM(r)\Delta m}{r^2} - A\Delta P. \tag{8.4}$$

The infinitesimal mass, Δm, is given by

$$\Delta m = \rho A \Delta r. \tag{8.5}$$

Substituting for Δm in the equation of motion, dividing by Δr, and taking the limit $\Delta r \to 0$, $\Delta P \to 0$, we obtain

$$\rho a = -\frac{GM(r)\rho}{r^2} - \frac{dP}{dr}, \tag{8.6}$$

Under equilibrium, the acceleration is zero. Hence we obtain

$$\frac{dP(r)}{dr} = -\frac{GM(r)\rho}{r^2} = -\rho g(r), \tag{8.7}$$

where we have defined the acceleration due to gravity $g(r) = GM(r)/r^2$. This equation gives us the rate of change of pressure as a function of r under equilibrium conditions. We notice that dP/dr is negative. This implies that at equilibrium, the pressure decreases with an increase in r. This is reasonable because the pressure at any point essentially balances the gravitational force acting on the column of gas on top. As we move to larger values of r, the column height gets smaller and the gas becomes less dense. Hence the gravitational force acting on it becomes smaller, resulting in a smaller pressure required to counter-balance it.

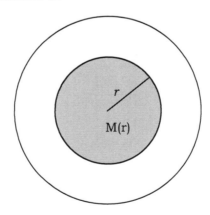

FIGURE 8.3: The variable, $M(r)$, is defined as the mass of the segment of radius r of a star.

8.2   Mass Distribution

We next obtain a differential equation for the mass $M(r)$ contained within the radius r (Figure 8.3). The infinitesimal mass contained within radius r and $r + \Delta r$ is given by

$$\Delta M(r) = 4\pi r^2 \rho \Delta r. \tag{8.8}$$

Taking the limit, $\Delta r \to 0$ and $\Delta M(r) \to 0$, we obtain

$$\frac{dM(r)}{dr} = 4\pi r^2 \rho. \tag{8.9}$$

This equation gives us the rate at which $M(r)$, the mass contained within a sphere of radius r, changes with r.

8.3   Energy Production

A star produces energy predominantly in its core due to nuclear fusion reactions. At steady state, the total luminosity of the star is equal to the rate of energy production. Let $L(r)$ be the total energy produced by the star within a sphere of radius r per unit time. At steady state, this is the luminosity or the rate at which energy flows outward from this sphere. We define the energy production coefficient, $\epsilon(r)$, to be the amount of energy released per unit time per unit mass. We can express the infinitesimal energy $\Delta L(r)$ released within the radius r to $r + \Delta r$ as

$$\Delta L(r) = \epsilon(r)\Delta M(r). \tag{8.10}$$

Substituting for $\Delta M(r)$, dividing by Δr, and taking the limit, $\Delta r \to 0$, we obtain

$$\frac{dL(r)}{dr} = 4\pi r^2 \rho \epsilon(r). \tag{8.11}$$

The total luminosity of a star is simply equal to $L(R)$, where R is the radius of the star. As we discuss later, the predominant source of energy in a star is nuclear fusion reactions. A detailed understanding of these reactions is necessary to compute the energy production coefficient ϵ. We address this subject in the next chapter.

8.4   Temperature Gradient

A star has a very high temperature in its core. The temperature drops as we go toward the surface. The energy produced in the core is transported to the surface either through radiative or convective transport. Radiative transport means that energy is transported directly by the propagation of electromagnetic radiation. For example, the energy from the Sun is transported to Earth by this mechanism. Inside stars, radiative transport dominates in regions of very high temperature, such as inside their cores. In the case of convective transport, energy is transported due to the motion of parcels of hot gas. For example, the gas near a flame gets heated and rises, leading to a convective transport of energy. In stellar interiors, this mechanism dominates in regions of lower temperature closer to the surface. The phenomenon of conductive transport, which plays a dominant role inside solids, does not contribute in the case of stars.

We next obtain a differential equation for dT/dr under equilibrium conditions, considering radiative and convective transport separately.

8.4.1   Radiative Transport

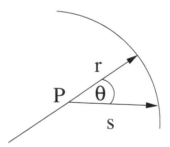

FIGURE 8.4: Radiation field at the point P. We consider radiation propagating in the direction s which makes an angle θ with respect to the radial direction.

Here we give a simplified treatment of radiative transport. Let's consider radiation at the point P propagating at an angle θ to the radial direction, as shown in Figure 8.4. The path taken by radiation is labeled by the symbol s. The rate at which the specific intensity I_λ changes is given by Equation 7.16. We are interested in applying this to the stellar interior where, to a good approximation, κ_λ and ρ depend only on the distance r from the center. Hence we obtain

$$\frac{dI_\lambda(r,\theta)}{ds} = -\kappa_\lambda(r)\rho(r)I_\lambda(r,\theta) + j_\lambda(r,\theta). \tag{8.12}$$

Here λ is the wavelength, κ_λ the absorption coefficient, ρ the density of the medium, and j_λ the emission coefficient of the medium. The specific intensity I_λ and emission coefficient j_λ also depend on the angle θ. The angular dependence of j_λ arises due to induced emission and is proportional to I_λ. A detailed discussion of this point is beyond the scope of the present book. Interested readers can refer to *Principles of Stellar Evolution and Nucleosynthesis* by D. D. Clayton. Here we simply note that due to this proportionality, we can absorb the θ-dependent part of j_λ in the term $\kappa_\lambda \rho I_\lambda$ with a redefinition of κ_λ. We denote the resulting absorption coefficient by the symbol κ'_λ and the isotropic part of j_λ by j'_λ. Hence j'_λ depends only on r. The resulting equation can be expressed as

$$\frac{dI_\lambda(r,\theta)}{ds} = -\kappa'_\lambda(r)\rho(r)I_\lambda(r,\theta) + j'_\lambda(r). \tag{8.13}$$

At any point, the star is locally in thermal equilibrium at temperature $T(r)$. Hence, to a good approximation, the specific intensity $I_\lambda(r,\theta)$ is equal to the blackbody intensity, $B_\lambda(r)$, corresponding to temperature $T(r)$. However, a star is not in exact thermal equilibrium. There is a net outward flux of radiation and hence $I_\lambda(r,\theta)$ deviates from the blackbody distribution. This deviation is important in the present derivation. In terms of r, we can express Equation 8.13 as

$$\cos\theta \frac{1}{\kappa'_\lambda(r)} \frac{\partial I_\lambda(r,\theta)}{\partial r} = -\rho(r)I_\lambda(r,\theta) + \frac{j'_\lambda(r)}{\kappa'_\lambda(r)}. \tag{8.14}$$

Here we have used $\partial r/\partial s = \cos\theta$ and divided throughout by κ'_λ. Furthermore, on the left-hand side of this equation, we have used the approximation $I_\lambda(r,\theta) \approx B_\lambda(r)$, the blackbody intensity, and ignored the θ dependence of $I_\lambda(r,\theta)$. We next integrate this equation over λ and set

$$I = \int_0^\infty d\lambda I_\lambda, \tag{8.15}$$

$$\frac{1}{\bar{\kappa}(r)} \int_0^\infty d\lambda \frac{dI_\lambda}{dr} = \int_0^\infty d\lambda \frac{1}{\kappa'_\lambda(r)} \frac{dI_\lambda}{dr}, \tag{8.16}$$

$$J(r) = \int_0^\infty d\lambda \, (j'_\lambda(r)/\kappa'_\lambda(r)). \tag{8.17}$$

The mean $\bar{\kappa}(r)$ is called the Rosseland mean opacity. We also have

$$\int_0^\infty d\lambda \frac{dI_\lambda}{dr} = \frac{d}{dr} \int_0^\infty d\lambda \, I_\lambda = \frac{dI}{dr}. \tag{8.18}$$

This gives us

$$\frac{\cos\theta}{\bar{\kappa}(r)} \frac{dI}{dr} = -\rho I + J(r). \tag{8.19}$$

We emphasize that we are neglecting the θ dependence of I on the left-hand side of this equation. We next multiply this equation by $\cos\theta$ and integrate over the entire solid angle, to obtain

$$\frac{1}{\bar{\kappa}(r)}\int d\Omega\cos^2\theta\,\frac{dI}{dr} = -\rho\int d\Omega\cos\theta\,I + \int d\Omega\cos\theta\,J\,, \qquad (8.20)$$

where $d\Omega = \sin\theta d\theta d\phi$. Because J depends only on r, we find that

$$\int d\Omega\cos\theta\,J = 0\,. \qquad (8.21)$$

In the limit of exact blackbody, $I = B$, where B is the blackbody intensity corresponding to the temperature at radial distance r. In that case, even the first term on the right-hand side of Equation 8.20 would vanish. However, this term is non-zero because the intensity is not exactly isotropic at any point and leads to an outward flux of radiation. We, therefore, set

$$\int d\Omega\cos\theta\,I = F(r)\,, \qquad (8.22)$$

where $F(r)$ is the flux density at radius r. The flux density is related to the total radiation flux or luminosity $L(r)$ at radius r by

$$F(r) = \frac{L(r)}{4\pi r^2}\,. \qquad (8.23)$$

We caution the reader that $F(r) \neq \sigma T^4(r)$. However, at the surface, the radiative flux $F(R) = \sigma T^4(R)$. On the left-hand side of Equation 8.20, the dominant contribution is obtained by setting $I = B(r)$, where $B(r)$ is the blackbody intensity at temperature $T(r)$. Setting $B(r) = \sigma T^4(r)/\pi$ and taking its derivative, we find

$$\frac{16\sigma}{3\bar{\kappa}}T^3\frac{dT}{dr} = -\rho F(r)\,. \qquad (8.24)$$

We set the Stefan-Boltzmann constant $\sigma = ca/4$, where a is the radiation constant and express $F(r)$ in terms of $L(r)$ to obtain

$$\frac{dT(r)}{dr} = -\frac{3\bar{\kappa}\rho L}{16\pi r^2 caT^3}\,. \qquad (8.25)$$

This gives the equation for the temperature gradient in regions where radiative transport dominates.

8.4.2 Convective Transport

We next obtain the equation of the temperature gradient assuming that convection provides the dominant contribution. Consider a small parcel of gas inside the stellar medium, as shown in Figure 8.5. If its temperature T' is higher than the ambient temperature T, then it will have a lower density. The surrounding gas exerts a force on it in the outward direction. Hence the gas parcel will tend to rise. If it is cooler, then it will tend to fall. This process of the rise and fall of the gas parcel can be considered adiabatic, to a very good approximation. This means that during this process, the gas parcel does not exchange heat with the surrounding medium. This is due to the fact that the time scale over which it exchanges heat with the surroundings is much larger than that of the rise and fall of the parcel. Let P' and T' be the pressure and temperature of the gas inside the parcel. As the parcel rises or falls, the change in pressure and temperature follows the standard adiabatic gas law,

$$P'^{1-\gamma}T'^{\gamma} = constant\,, \tag{8.26}$$

where $\gamma = C_P/C_V$ is the gas constant. Here C_V and C_P are, respectively, the specific heat capacities at constant volume and pressure.

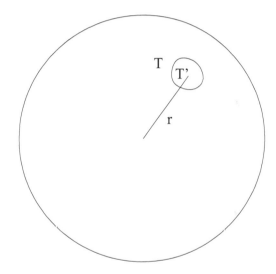

FIGURE 8.5: A small parcel of gas at distance r from the center. The temperature inside the parcel is denoted by T'. The ambient temperature is $T(r)$.

Let's assume that initially the parcel has the same temperature as the surroundings. Due to a small perturbation, it starts to rise. As the parcel rises, its temperature decreases according to the adiabatic law, Equation 8.26. Note that the pressure and density inside a star decreases with increase in r. If the temperature of the parcel falls with r at a rate less than that of the surroundings, that is, if

$$\left|\frac{dT'}{dr}\right| < \left|\frac{dT}{dr}\right|\,, \tag{8.27}$$

then it will be unstable and continue to rise. This is because in this case, it will continue to encounter cooler environment as it rises. In contrast, if its temperature falls at a rate faster than the surroundings, that is, if

$$\left|\frac{dT'}{dr}\right| > \left|\frac{dT}{dr}\right|, \tag{8.28}$$

then it will be stable and will start oscillating about its mean position. In this case, convection cannot occur. If convection dominates energy transport, then the gas will be thoroughly mixed due to its rise and fall. In this case, in equilibrium, the ambient temperature gradient would be the same as the adiabatic temperature gradient. This can be understood as follows: If the temperature gradient is smaller than this, then convection cannot occur, in contrast to our assumption that it dominates the energy transport. If the temperature gradient is higher than this, then gas will rise and fall until it becomes equal to the adiabatic gradient.

Using the adiabatic law, the temperature of the gas can be expressed as

$$T = CP^{1-1/\gamma},$$

where C is a constant. By differentiating with respect to P and eliminating C, we find

$$\frac{dT}{dr} = \left(1 - \frac{1}{\gamma}\right)\frac{T}{P}\frac{dP}{dr}, \tag{8.29}$$

which is the required equation for the temperature gradient. It is valid when convection dominates energy transport.

8.5 Boundary Conditions

In order to obtain a unique solution to the stellar differential equations derived in this chapter, we need to specify suitable boundary conditions. These are given as follows:

At $r = R$, where R is the radius of the star,

$$M(R) = M_0, \tag{8.30}$$

where M_0 is the total mass of the star. Furthermore, it is convenient to set the temperature and pressure at the surface equal to zero, that is, $T(R) = 0$ and $P(R) = 0$. This is not entirely true for real stars where this value is small but non-zero. However, this small value is not expected to make a major difference in the final results.

At $r = 0$, one sets the boundary conditions $M(0) = 0$ and $L(0) = 0$. This is equivalent to the assumption that there is no hard core at the center of the star. The mass varies smoothly as we go to the center.

Besides these boundary conditions, we also need to give the equation of state and the chemical composition of the medium. The equation of state specifies the relationship between the pressure and temperature and is derived later in this chapter. As we will see, the equation of state for gases, in which the particles move at nonrelativistic speeds, is very different in comparison to radiation.

8.6   Rosseland Mean Opacity

A star is a gaseous medium, and is opaque to photons throughout its interior. The energy of a star is generated in the thermonuclear core. The photons produced in the core get absorbed and scattered in the medium. The medium is at steady state. Hence the energy gained due to absorption is simultaneously being lost by the emission of photons. The net effect of these processes is that the energy per photon decreases with distance from the center. However, by energy conservation, the total number of photons increases. As we approach the surface, the opacity becomes sufficiently small that photons can propagate freely.

The radiative transport, or propagation, of photons through stellar interiors, was discussed in Section 8.4.1. The Rosseland mean opacity, $\bar{\kappa}(r)$, defined in Equation 8.17 governs the propagation of radiation, after averaging over wavelengths. The main contributions to opacity were already discussed in Chapter 7. We have several processes that lead to absorption of a continuous range of frequencies. These are (1) inverse Bremsstrahlung or free-free absorption, (2) bound-free transitions or photo-ionization, and (3) Compton scattering. We denote the corresponding opacities as $\kappa_{\lambda,ff}$, $\kappa_{\lambda,bf}$, and $\kappa_{\lambda,cs}$, respectively. The bound-free process contributes to a continuous range of photon frequencies, greater than the ionization frequency. The contribution of the Compton scattering process is relatively small compared to other processes at most temperatures. It contributes only at very high temperature, where the bound electron density is small and most electrons are in the free state. In the limit $h\nu << m_e c^2$, the scattering cross section is given by

$$\sigma_T = \frac{8\pi}{3}\left(\frac{e^2}{m_e c^2}\right)^2 . \tag{8.31}$$

This is called the Thomson cross section for the scattering of photons with electrons. As an exercise, you can show that its value is equal to 0.665×10^{-28} m^2. Note that this cross section is independent of frequency. This implies that the corresponding opacity $\kappa_{\lambda,cs}$ is independent of frequency. We also get a contribution to opacity at discrete frequencies due to bound-bound transitions. Here a bound electron absorbs a photon and makes a transition to another bound state. The frequency of the photon ν is given by $h\nu = \Delta E$, where ΔE is the energy difference between the two states. We denote the corresponding opacity by $\kappa_{\lambda,bb}$. The Rosseland mean is obtained after taking the sum of all these contributions, that is,

$$\kappa_\lambda = \kappa_{\lambda,bb} + \kappa_{\lambda,bf} + \kappa_{\lambda,ff} + \kappa_{\lambda,es}.$$

The mean opacity increases with an increase in density ρ. It also depends on the composition of the medium. For fixed density and composition, it first increases with temperature, reaches a maximum, and then starts to decrease. In Figure 8.6 we show the mean opacity as a function of temperature for fixed $R = \rho/T_6^3$. In this plot, the mass ratios $X = 0.7$ and $Y = 0.28$ correspond to conditions inside the solar interior. It is convenient to choose a fixed value of R because in stellar interiors, ρ is approximately proportional to T^3. At very high temperatures $\bar{\kappa}$ approaches a constant value. At low temperatures, we only have neutral atoms, predominantly hydrogen and helium, in the medium, and opacity is dominated by bound-bound and bound-free transitions. As the temperature increases, the free electron density starts to increase. This contributes to free-free transition leading to a rise in opacity with temperature.

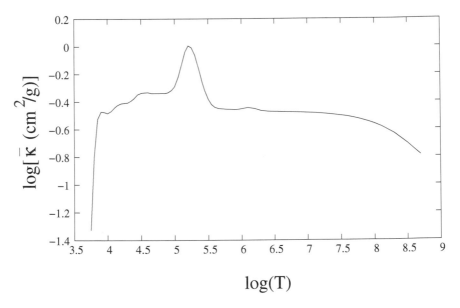

FIGURE 8.6: The logarithm of the Rosseland mean opacity, $\bar{\kappa}$ (cm^2/g), as a function of temperature $\log(T)$ for parameters $X = 0.7$, $Y = 0.28$ and $\log(R) = \log(\rho/T_6^3) = -6$. Here the temperature is in Kelvin, density ρ is in g/cm^3 and $T_6 = T/(10^6$ K$)$. The mass fractions X and Y correspond to conditions inside the solar interior beyond the core. The data are taken from OPAL opacity tables, C. A. Iglesias and F. J. Rogers, *Astrophysical Journal* 464, 943 (1996).

At sufficiently high temperatures, most of the hydrogen atoms are ionized. Hence the electron density reaches saturation beyond a certain temperature. Furthermore, the mean free-free opacity decreases with temperature. Hence the Rosseland mean opacity reaches a maximum and starts to decrease beyond a certain temperature. Eventually at very high temperatures, the electron scattering process dominates. The mean opacity becomes roughly independent of temperature. This is essentially due to the wavelength independence of the Thomson scattering cross section.

As explained above, radiative transport from the core of the star to the surface happens through repeated scattering, absorption, and emission processes. A high-energy photon in the core, produced by nuclear reactions, may scatter to produce a lower energy photon, transferring part of its energy to an electron. The electron may lose the additional energy by a process such as bremsstrahlung to produce another photon. The net result is that we have a greater number of photons, each of lower energy in comparison to the original photon. Hence as we move outward from the center, the energy per photon, or equivalently the radiation blackbody temperature, decreases.

The temperature and density of a star decrease with increasing radius. Hence in different regions inside a star, different processes contribute to opacity. In the cores of stars, the temperature is sufficiently high that the electron scattering process gives the dominant contribution to opacity. As we move to larger radius, the opacity of a star typically increases, with a larger contribution coming from free-free, bound-free, and bound-bound transitions. For main sequence stars of mass close to solar mass, radiative transport dominates at small radius. However, for a large radius, the opacity becomes very large, which causes very large temperature gradients. At this point, the medium becomes unstable to convection, which starts to dominate energy transport.

Eventually, as we approach the visible surface of a star, the opacity becomes very small. Correspondingly, the mean free path of photons $l_\lambda = 1/(\kappa_\lambda \rho)$ becomes very large, and photons are able to propagate freely. The attenuation of radiation due to absorption and scattering is governed by Equation 7.15. We can express it as

$$I(s) = I_0 e^{-\int ds/l_\lambda} . \tag{8.32}$$

For a uniform medium, $I(\Delta s) = I_0 e^{-\Delta s/l_\lambda}$, where Δs is the distance of propagation.

Due to free propagation near the surface, the conditions for blackbody distribution are violated. The radiation is no longer in thermal equilibrium. Just below the surface, the medium is opaque and the radiation remains in thermal equilibrium. In this region, the dominant contribution to continuum opacity, that is, opacity for a continuous range of frequencies, depends on the nature of the star. For very hot O type stars, the electron scattering process dominates the continuum opacity. For A and B type stars, the dominant contribution comes from photo-ionization of hydrogen and free-free absorption. For other stars, including the Sun, the cause of opacity near the surface remained a puzzle for a long time due to their relatively low surface temperatures. At these temperatures, the neutral hydrogen and helium atoms are predominantly in their ground state, requiring frequencies in the ultraviolet range for ionization. Hence these atoms cannot contribute to continuum opacity at visible frequencies, which dominate the spectrum of such stars. This problem was eventually solved by postulating the presence of H^- ion in the atmospheres of these stars. The H^- ion has two electrons bound to a proton nucleus (see Figure 8.7). Its ionization energy is 0.754 eV. Hence it can be ionized by a photon at visible or larger frequencies and thus gives rise to continuum opacity in this range.

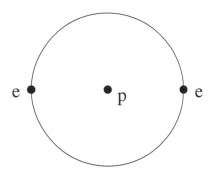

FIGURE 8.7: The H^- ion gives dominant contribution to solar opacity near the surface.

8.7   Equation of State

We have so far obtained the equations for the pressure, mass, luminosity, and temperature gradient for a star in equilibrium. In order to complete the system of equations, we need to specify the equation of state, that is, the relationship between the pressure and temperature. We next obtain the equation of state for radiation and gas, the two main components of stars.

Consider gas at temperature T contained inside a cubical box, which is at rest. The individual gas particles have non-zero velocities and undergo motions in random directions. Let us denote the velocity of the ith particle by $\vec{v}_i$. Because the box is at rest, the mean velocity $\langle \vec{v} \rangle$ of all the particles is zero, that is,

$$\langle \vec{v} \rangle = \frac{1}{N} \sum_i \vec{v}_i = 0 \,, \tag{8.33}$$

where N is the total number of particles. The mean value of $\vec{v} \cdot \vec{v} = v^2$, which is equal to the square of the speed of the particle, however, is not zero. By the kinetic theory of gases, this is related to the temperature of the gas. Hence the higher the temperature, the faster the motion of the particles. Notice that it is the random motion of the particles that determines the temperature. Mathematically, the particle velocities follow the Maxwell–Boltzmann distribution, discussed in the Appendix (Section 8.9).

To a good approximation we can assume the gas is ideal, that is, interactions between particles are small. The gas particles are assumed to be free and they occasionally undergo elastic collisions with other particles and with the walls of the box. In an elastic collision, the mechanical energy remains conserved, that is, the sum of the kinetic energies of all the particles remains the same before and after collision. Due to collisions, the particles also exert pressure on the walls.

We first derive the equation of state for a gas consisting of particles moving at nonrelativisitic velocities, that is, speeds much smaller than the speed of light. Later we generalize it to relativistic particles, such as photons.

8.7.1 Ideal Gas Law

Let us first consider a gas consisting of a single species. This means that all the gas particles are identical. These particles may be hydrogen molecules (H_2), oxygen molecules (O_2), nitrogen molecules (N_2), etc. The equation of state for such a gas is given by the formula

$$P = \frac{N}{V}kT,$$

(8.34)

where P, T, N, and V, respectively, denote the pressure, temperature, number of particles, and volume of the gas, and k is the Boltzmann constant. We derive this basic equation later in this section.

We need to apply the gas law, Equation 8.34, for gases containing many different species. For example, our atmosphere contains N_2, O_2, and other species, such as water vapor, carbon dioxide, in small proportions. The stellar medium also consists of many different species. In many cases, particularly in the interior of stars, the temperature is so high that all atoms are typically found in their ionized state. In this case, the gas consists of ions and electrons. We next consider an ideal gas consisting of several different species such as hydrogen, helium, and heavier elements. The heavier elements with atomic number $Z > 2$ are collectively referred to as metals in Astronomy. We assume that all these species are in thermal equilibrium at temperature T and occupy a volume V. Each species exerts a partial pressure P_i, which is given by

$$P_i = \frac{N_i}{V}kT.$$

(8.35)

Here N_i is the total number of particles of species i. Let P be the total pressure of all the particles in the gas. This is equal to the sum of the partial pressures exerted by each species. Similarly, N, the total number of particles of all types, including ions and electrons, is equal to the sum over individual species. We can write P and N as

$$P = \sum_i P_i,$$

$$N = \sum_i N_i, \qquad (8.36)$$

where the sum is over all the particle species. It is clear that P and N also satisfy Equation 8.34. It is convenient to introduce the mean molecular weight of the gas, defined by

$$\mu = \frac{M}{Nm_H}, \qquad (8.37)$$

where M is the total mass of the gas and m_H is the mass of a hydrogen atom. In terms of μ, the gas law can be written as

$$P = \frac{k}{\mu m_H} \rho T, \qquad (8.38)$$

where $\rho = M/V$ is the density of the gas. We next derive the gas law and later generalize it to obtain the equation of state for radiation.

8.7.1.1 Derivation of the Ideal Gas Law

Consider a box of sides Δx, Δy, and Δz. The coordinate system is aligned such that the axes are parallel to the three walls of the box, as shown in Figure 8.8. Let N be the total number of gas particles in the box. We are interested in obtaining the pressure exerted by the gas on the walls of the box. The gas particles continuously hit the wall, undergoing elastic collisions. Consider a wall aligned along the y-z plane, perpendicular to the x-axis. A particle moving along the x-axis undergoes a head-on elastic collision with the wall, as shown in Figure 8.9(a). During collision, the wall exerts a sudden or impulsive force on the particle in the $-x$ direction. This force acts over a very short time but has very high magnitude. Because the collision is assumed to be elastic, the kinetic energy, $mv^2/2$, of the particle must remain conserved in this collision. Hence its speed remains unchanged. The particle simply bounces back with the same speed. If the initial momentum of the particle is $p_x = mv_x$, the final momentum is $-p_x$. In general, the particle hits the wall at some angle, as shown in Figure 8.9(b). We can analyze this collision by considering the velocity components in the three directions independently. The x component of the velocity again reverses direction, as before. However, the y and z components remain unchanged because the wall exerts an impulsive force only in the x direction. The x component of its momentum, p_x, changes to $-p_x$. Hence, the change in this component is

$$\Delta p_x = 2p_x,$$

with the remaining two components, p_y and p_z, unchanged.

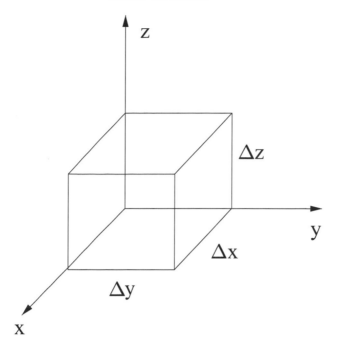

FIGURE 8.8: A box of sides Δx, Δy, and Δz containing gas at temperature T and pressure P.

The particle returns to the wall after a time

$$\Delta t = 2\Delta x/v_x.$$

This is easily understood. Ignore the motion in y and z directions. As far as the motion in the x direction is concerned, the particle simply keeps bouncing back and forth between the two opposite walls. Hence it travels a distance $2\Delta x$ along the x direction between two consecutive collisions with a particular wall. Here we have assumed that the particle does not collide with another particle. It collides only with the walls of the container. This assumption is valid for a gas at low density. The pressure exerted on the wall is given by

$$P = \frac{F}{A} = \frac{\sum \Delta p_x/\Delta t}{\Delta y \Delta z} = \frac{\sum p_x v_x}{\Delta x \Delta y \Delta z},$$

where $A = \Delta y \Delta z$ is the area of the wall, and the sum in the numerator is over all the particles in the box. We can express the pressure as

$$P = \frac{N < p_x v_x >}{V}, \tag{8.39}$$

where V is the volume of the box, N is the total number of gas particles, and $< p_x v_x >$ is the mean value of the product $p_x v_x$ over the gas particles.

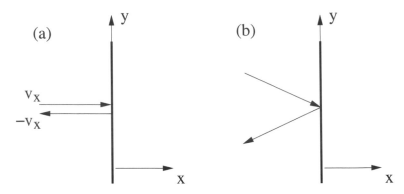

FIGURE 8.9: A particle undergoing elastic collision with a wall in the y-z plane. The z component points outward, perpendicular to the plane of the paper. (a) The incoming particle, moving in the $+x$ direction, undergoes a head-on elastic collision with the wall. After collision, the particle reverses its direction, while its speed remains unchanged. (b) The particle is incident at an angle. After collision, the x component of the velocity of the particle reverses its direction, while its remaining components remain unchanged.

We next write the momentum of the gas particles, assumed to be moving at nonrelativistic speeds ($v \ll c$), as $p_x = mv_x$ to get

$$P = \frac{Nm < v_x^2 >}{V} .$$

Assuming isotropy, that is, all directions are equivalent, we find $< v_x^2 >=< v_y^2 >=< v_z^2 >=< v^2 > /3$. The last step follows because $v^2 = v_x^2 + v_y^2 + v_z^2$. We may also perform the angular averaging directly by using spherical polar coordinates. We have, for example,

$$< v_z^2 >=< v^2 \cos^2 \theta >=< v^2 > \frac{\int_0^\pi d\theta \sin \theta \cos^2 \theta}{\int_0^\pi d\theta \sin \theta} =< v^2 > /3, \qquad (8.40)$$

where $\langle v^2 \rangle$ represents the thermal average of v^2. Therefore the pressure is given by

$$P = \frac{Nm < v^2 >}{3V} . \qquad (8.41)$$

We assume that the speeds of the gas particles follow the Maxwell–Boltzmann distribution. Using the standard result from the kinetic theory of gases (see Section 8.9),

$$< v^2 >= \frac{3kT}{m} ,$$

we obtain the standard gas equation of state,

$$P = \frac{NkT}{V} . \qquad (8.42)$$

We can also express this in terms of the energy of the gas by using $E = N < \epsilon >= Nm < v^2 > /2$, where $< \epsilon >$ is the mean value of energy per particle. We obtain

$$P = \frac{2E}{3V} .$$ (8.43)

We next determine the mean molecular weight, μ, defined in Equation 8.37, in terms of the relative mass abundance of different elements in the star. Let $M = \rho V$ be the total mass of the gas in volume V. Let X, Y, and Z' respectively, denote the relative mass abundance of hydrogen, helium, and metals (elements with $Z > 2$). It is clear that

$$X + Y + Z' = 1 .$$

In the stellar interiors, all hydrogen and helium atoms are ionized. Hence each hydrogen atom contributes two particles to the gas, an electron and a proton. It also contributes a unit mass, in atomic mass units. Similarly, a helium atom contributes three particles, the nucleus and two electrons, and has mass equal to 4. Let us denote the number densities of hydrogen and helium nuclei by N_H, N_{He}, respectively. The number of particles contributed to the medium by hydrogen and helium are, therefore, $2N_H$ and $3N_{He}$. We ignore the contribution due to metals, whose contribution is small and will be included later. The total number of particles can be expressed as

$$N = 2N_H + 3N_{He} .$$

The relative mass abundance of hydrogen and helium is given by

$$X = \frac{N_H m_H}{M} ,$$ (8.44)

$$Y = \frac{4N_{He} m_H}{M} .$$ (8.45)

Note that the mass of each hydrogen atom is approximately m_H and that of helium atom is $4m_H$. Here we neglect the nuclear and atomic binding energies. Hence we find

$$N = \frac{M}{m_H}(2X + 3Y/4) .$$

This leads to the mean molecular weight:

$$\mu = \frac{M}{Nm_H} = \frac{1}{2X + 3Y/4} .$$ (8.46)

We next compute the contribution due to a heavier element. Consider an atom with atomic number Z and weight A. Let us assume that it is partially ionized and contributes $\bar{Z}$ particles to the gas. If we assume that it is fully ionized, then $\bar{Z} = Z + 1$. Let N_A be the number density of these ions and X_A

its relative mass abundance. The mass fraction due to a heavy element can be expressed as

$$X_A = \frac{N_A A m_H}{M} .$$

This finally gives the formula for the mean molecular weight, including the contribution from a heavier element, as

$$\mu = \frac{1}{2X + 3Y/4 + \bar{Z} X_A/A} . \tag{8.47}$$

The contribution from all the heavier elements can be included in this manner.

8.7.1.2   Radiation Pressure

We next obtain an expression for pressure exerted by a photon gas. In this case the speed of photons, $v = c$. We assume that the photon frequencies follow the blackbody distribution at temperature T. The energy and momentum of a particle of mass m moving at velocity close to c is given by the special theory of relativity. These can be written as

$$\epsilon = m\gamma c^2 , \tag{8.48}$$

$$\vec{p} = m\gamma \vec{v} , \tag{8.49}$$

respectively, where $\gamma = 1/\sqrt{1 - v^2/c^2}$. Eliminating m from these equations we obtain

$$\vec{p} = \frac{\epsilon}{c^2} \vec{v} . \tag{8.50}$$

The equations for ϵ and $\vec{p}$, Equations 8.48 and 8.49, respectively, break down in the limit of zero mass and $v \to c$. This is because $\gamma \to \infty$ as $v \to c$. However, the relationship between ϵ and $\vec{p}$, Equation 8.50, is valid for all particles, including photons. It also implies that the energy of a relativistic particle, $\epsilon = pc$, where p is its total momentum.

Assuming isotropy, we have

$$< p_x v_x > = < p_y v_y > = < p_z v_z > = \frac{1}{3} < \vec{p} \cdot \vec{v} > = \frac{\langle \epsilon \rangle}{3} . \tag{8.51}$$

We also have, for example,

$$< p_z v_z > = \left\langle \frac{\epsilon}{c^2} v_z v_z \right\rangle = \left\langle \epsilon \cos^2 \theta \right\rangle = \frac{\langle \epsilon \rangle}{3} . \tag{8.52}$$

Substituting in Equation 8.39 we obtain

$$P = \frac{N \langle \epsilon \rangle}{3V} = \frac{E}{3V} = \frac{4\pi}{3c} I , \tag{8.53}$$

where I is the intensity of radiation and we have used the formula for the energy density of the blackbody $u = E/V = 4\pi I/c$. The blackbody intensity is given by $I = \sigma T^4/\pi$, where σ is the Stefan–Boltzmann constant. We eventually find the pressure

$$P = \frac{1}{3}aT^4 \,, \tag{8.54}$$

where $a = 4\sigma/c$.

8.8 Energy Production in Stars

The stars produce energy through nuclear fusion in their cores. This is now well known and observationally verified. However, this question had puzzled scientists for a long time. At the time when the phenomenon of nuclear fusion was unknown, scientists tried to explain the solar luminosity by the known energy production mechanisms at that time. These included the release of gravitational and chemical energies. Gravitational potential energy is released as the star slowly contracts during its life cycle. One can determine the total amount of gravitational potential energy released by the Sun so far. This can then be compared to the total energy radiated by the Sun. However, in order to do so, we need to know the lifetime of the Sun and also its luminosity during this period. Historically, in the nineteenth century, these quantities were not well known. Hence scientists assumed that the source of stellar energy may be gravitational potential energy or chemical reactions and thus computed the lifetime of the Sun based on this hypothesis.

Let's first determine the total amount of energy that can be released due to gravitational potential energy. Consider a star of radius R, mass M, and density ρ. The potential energy of a spherical shell of radius r, thickness dr, and mass dm is $dU_g = -GM(r)dm/r$. Here, $M(r)$ is the mass contained within radius r of the star and $dm = 4\pi r^2 dr\rho$. Hence the total potential energy of a star of radius R and density ρ is

$$U_g = -4\pi G \int_0^R dr M(r)\rho r \,. \tag{8.55}$$

In order to perform this integral, we need the dependence of ρ on r. Here we are only interested in an order of magnitude estimate. By assuming a constant ρ and ignoring the overall dimensionless factor, we obtain

$$U_g \sim -\frac{GM^2}{R} \,. \tag{8.56}$$

We next need to determine the total mechanical energy of a star, given by $E = K + U_g$, where K denotes the kinetic energy. In order to do so, we use the virial theorem, as derived in Chapter 5. The theorem, Equation 5.23, states that, for a system in equilibrium, $-2 < K >=< U >$, where $< x >$

denotes the average of the quantity x over time. Hence the total mechanical energy of a star is

$$E =< K > + < U >= \frac{1}{2} < U > . \tag{8.57}$$

Using the formula for gravitational potential energy, we find

$$E \sim -\frac{1}{2} \frac{GM^2}{R} . \tag{8.58}$$

Let's apply this to the Sun, assuming that initially its radius was extremely large. Hence its energy at that time was zero. Its final energy is

$$E_f \sim -\frac{1}{2} \frac{GM_S^2}{R_S} ,$$

where M_S is its mass and R_S its radius today. Hence the total gravitational energy released by the Sun over its lifetime is

$$\Delta E_g \approx \frac{1}{2} \frac{GM_S^2}{R_S} \approx 10^{48} \text{ ergs} . \tag{8.59}$$

In order to compare with observations, we can assume that the luminosity of the Sun was roughly uniform throughout its lifetime and equal to its luminosity today ($L_S = 3.9 \times 10^{33}$ ergs/sec). With this assumption we obtain an estimate of the lifetime of the Sun,

$$\Delta t = t_{\text{th}} = \frac{\Delta E_g}{L_S} \approx 10^7 \text{ years} , \tag{8.60}$$

where we have included only the gravitational energy. This is called the thermal or Kelvin-Helmholtz time scale in honor of the scientists who proposed this mechanism for solar luminosity. The corresponding time scale of the Sun, assuming that the chemical energy dominates, turns out to be smaller in comparison to the Kelvin-Helmholtz time scale. Hence, before the discovery of nuclear reactions, this was believed to be a reasonable estimate of the lifetime of the Sun.

There now exists considerable evidence that t_{th} is much lower than the actual lifetime of the Sun. One can use radioactive dating to determine the age of lunar rocks or meteorites. This gives an age on the order of 4×10^9 years. The rocks on Earth give a much lower value, perhaps because the earlier rocks may have been destroyed. In order to compare this age with the time scale t_{th}, one has to assume that the luminosity of the Sun has not changed significantly over most of its lifetime. This can be justified by considering geological and biological activity, which requires the current luminosity of the Sun over a time period larger than 10^9 years. Incidentally, the earlier theoretical estimate of the lifetime of the Sun, t_{th}, was used as evidence against Darwin's theory of biological evolution. Later it was found that Darwin's theory is consistent with observations if the Sun is powered by nuclear energy.

We next make an estimate of the energy generated in the Sun through nuclear fusion. The basic fusion reaction that occurs inside the Sun's core is the fusion of hydrogen to form helium. There are several different processes or chains that contribute to this reaction. One of the dominant chains that contributes can be represented as

$$4\,^1\mathrm{H} \rightarrow\,^4\mathrm{He} + 2e^+ + 2\nu_e + 2\gamma\,,$$

where e^+, ν_e, and γ represent a positron, neutrino, and a photon, respectively. Here $^A\mathrm{X}$ denotes the nucleus X with mass number A. The reaction basically converts four hydrogen nuclei into a helium nucleus. During this process about 0.7% of the rest energy of the nuclei is released. We will make a detailed estimate of this energy later. Here we make a rough estimate by assuming that all the hydrogen in the inner core of the Sun, which contains about 10% of the solar mass, converts to helium. Hence the energy released is

$$E_{\mathrm{nuclear}} \approx 0.1 \times 0.007 \times M_S c^2 \approx 10^{51}\ \mathrm{ergs}\,. \tag{8.61}$$

This gives an estimate of the lifetime of the Sun, called the nuclear time scale, of

$$t_{\mathrm{nuclear}} = \frac{E_{\mathrm{nuclear}}}{L_S} \approx 10^{10}\ \mathrm{years}\,. \tag{8.62}$$

Hence we find an estimate that is consistent with all observational evidence for the lifetime of the Sun. This provides considerable evidence that the source of solar energy is the nuclear fusion reactions in its core. We consider stellar nuclear reactions in more detail in the next chapter.

8.9 Appendix: Maxwell–Boltzmann Distribution

Consider an ideal gas at temperature T. The probability of finding particles with kinetic energy lying between energy E and $E + dE$ is given by

$$f(E)dE = f_0 e^{-E/kT} dE\,, \tag{8.63}$$

where k is the Boltzmann constant and f_0 is the normalization. This is the Maxwell–Boltzmann distribution. The normalization is chosen such that

$$\int_0^\infty f(E)dE = 1\,. \tag{8.64}$$

Let m be the mass of gas particles. The kinetic energy of a particle moving with velocity $\vec{v}$ is $mv^2/2$, $v^2 = v_x^2 + v_y^2 + v_z^2$. In terms of velocity we can write the distribution as

$$f(v_x, v_y, v_z)d^3v = f_0 e^{-mv^2/2kT} d^3v\,. \tag{8.65}$$

Integrating, we obtain

$$1 = \int_{-\infty}^{\infty} \int_{-\infty}^{\infty} \int_{-\infty}^{\infty} f(v_x, v_y, v_z) dv_x dv_y dv_z = f_0 \left(\frac{2\pi kT}{m} \right)^{3/2}. \tag{8.66}$$

Hence, the distribution can be expressed as

$$f(v_x, v_y, v_z) d^3 v = \left(\frac{m}{2\pi kT} \right)^{3/2} e^{-mv^2/2kT} d^3 v. \tag{8.67}$$

Let N be the total number of particles in the gas in volume V. Hence the number of particles in this volume having a velocity between (v_x, v_y, v_z) and $(v_x + dv_x, v_y + dv_y, v_z + dv_z)$ is equal to $N f(v_x, v_y, v_z) d^3 v$. We can now compute the mean square velocity, $\langle v^2 \rangle$, of a gas particle. This is given by

$$\langle v^2 \rangle = \int d^3 v v^2 f(v_x, v_y, v_z), \tag{8.68}$$

where the integral is over the entire range of velocities. We obtain

$$\langle v^2 \rangle = \frac{3kT}{m}. \tag{8.69}$$

Exercises

8.1 Use the equation for hydrostatic equilibrium, Equation 8.7, to determine the pressure as a function of height for the Earth's atmosphere. In this case, we can neglect the curvature of the Earth and the equation becomes

$$\frac{dP}{dz} = -\rho g,$$

where z is the vertical distance above the surface. For small z, we can treat g as constant. Eliminate ρ from this equation by using Equation 8.38. Show that the solution can be written as

$$P(z) = P_0 \exp\left[-\frac{g\mu m_H}{k} \int_0^z \frac{dz'}{T(z')} \right],$$

where P_0 is the pressure at the surface of the Earth. As a very rough approximation for small z ($z < 100$ Km), assume that $T(z) = T_0 - \beta z$, where $T_0 = 300$K and $\beta = 1$K/Km, that is, the temperature changes by 1K per Km. Determine $P(z)$ in this case.

8.2 The pressure, $P(z)$ near the Earth's surface ($z < 100$ Km) can be approximated as

$$P(z) \approx P_0 e^{-z/z_0},$$

where z is the altitude, $z_0 \approx 7.3$ Km, and $P_0 \approx 101,000$ Pa ($Pa =$

N/m^2). Assume that in some region the ambient temperature at the surface is $T_0 = 300K$ and decreases with height at the rate of 15K/Km. Determine whether the air is stable or unstable to convection in this region. Assume $\gamma = 7/5$, applicable for diatomic gases, N_2 and O_2.

8.3 Determine the numerical value of the Thomson cross section, using Equation 8.31. You can use the relationship

$$\alpha = \frac{e^2}{\hbar c}$$

to replace e^2 in terms of $\hbar$ and c. Note that $\alpha \approx 1/137$ is the fine structure constant.

8.4 Derive Equation 8.24 for the temperature gradient by following the steps explained in text.

8.5 Perform the integral in Eq. 8.55 to obtain U_g assuming that the density is constant, $\rho = M/(4\pi R^3/3)$. Verify that you obtain Equation 8.56 up to an overall factor of order one.

8.6 Verify the estimates of the gravitational potential energy and nuclear energy of the Sun, given in Equations 8.59 and 8.61, respectively.

8.7 Determine the mean molecular weight by including the contribution due to all metals. Assume that all metals are ionized and replace $\bar{Z} = Z+1$. Furthermore, assume that $Z + 1 \approx A/2$. With these approximations, show that

$$\mu = \frac{1}{2X + 3Y/4 + Z'/2}, \tag{8.70}$$

where Z' is now interpreted as the total mass fraction of all elements with $Z > 2$.

8.8 Perform the integrals in Equation 8.66. You can use

$$\int_{-\infty}^{\infty} dx\, e^{-x^2} = \sqrt{\pi}\,.$$

Also derive Equation 8.69 by performing the integrals in Equation 8.68.

Chapter 9

Stellar Nuclear Reactions

The possibility that nuclear fusion reactions may be the dominant source of the energy of stars was first suggested by Arthur Eddington in 1920. However, at that time, a detailed mechanism was unknown. Later, in 1928, George Gamow showed that nuclear reactions proceed through the quantum mechanical tunneling process. A comprehensive treatment of the two main reaction chains, the PP chain and the CNO cycle, which contribute to the main sequence stars, was provided by Hans Bethe in 1939. This work firmly established the validity of the fusion hypothesis.

The cores of main sequence stars consist predominantly of hydrogen and helium nuclei. Heavier nuclei as well as isotopes of hydrogen and helium are also present, but at very small densities. The energy production of the main sequence stars is dominated by fusion reactions that convert hydrogen into helium. The nuclear reaction proceeds predominantly by two-body collisions because the probability of three particles undergoing a nuclear reaction is extremely small. The PP chain consists of a series of reactions that effectively convert four hydrogen nuclei into a helium nuclei. The CNO cycle utilizes carbon, oxygen and nitrogen as catalysts to convert hydrogen into helium. Before discussing nuclear reactions, we review some basic concepts.

9.1 Fundamental Interactions

The fundamental forces or interactions that contribute in nuclear reactions are

1. Strong interactions

2. Electromagnetic interactions

3. Weak interactions

Besides gravity, these are the only known fundamental forces. Gravity is negligible compared to other forces in nuclear reactions.

The strong interaction is the force responsible for nuclear fusion. It acts only on particles such as protons, neutrons, and other exotic particles such as pions, kaons, etc. These particles are collectively called hadrons and are composed of fundamental particles called quarks. This force does not act on particles such as electrons and photons. The strong force between particles is very large if they are within a distance of approximately 1 fm (1 fm = 10^{-13} cm) from one another. At larger distances, the force dies off very rapidly. It is responsible for binding the atomic nucleus and hence is also referred to as the nuclear force.

The electromagnetic force acts on all charged particles. It also acts on neutral particles, such as atoms, if they are bound states of charged particles. The photon acts as a messenger of this force. Two particles interact electromagnetically by exchanging a photon. Due to this interaction, a particle can also emit or absorb a photon. The electromagnetic force is a long-range force. At short distances, on the order of 1 fm, it is very weak compared to the strong force. However, at larger distances, it dominates.

The weak force is responsible for processes such as β-decay, which converts a neutron into a proton or a proton into a neutron. A free neutron can convert to a proton by the reaction

$$n \to p^+ + e^- + \bar{\nu}_e \; . \tag{9.1}$$

Here the symbol $\bar{\nu}_e$ denotes an electron anti-neutrino. A free proton cannot decay into a neutron because its mass is smaller than that of a neutron. However, in a bound state or in the presence of other particles, it can convert to a neutron. The weak force has very short range. At the distance of 1 fm, relevant to nuclear reactions, it is very weak compared to strong and electromagnetic interactions. Furthermore, it decays rapidly with an increase in distance.

A reaction that proceeds by strong interactions has a much higher rate in comparison to reactions driven by electromagnetic or weak forces. The rate is smallest for a reaction driven by weak interactions. Hence a nuclear reaction that involves a weak force will proceed at a relatively slow rate in comparison to that which proceeds by a strong force. Let's consider, for example, the PP chain that starts with the fusion of two protons. A fusion reaction between two protons by means of nuclear force is not possible because there does not exist a stable isotope with $Z = 2$ and $A = 2$. The only possibility is to convert one of the protons into a neutron by the weak force and form a deuteron ^{2}H. Hence the rate of this reaction is extremely slow compared to subsequent reactions, which involve fusion of deuteron nuclei.

9.2   Fundamental Particles

The known fundamental particles are photons, gluons, W^+, W^-, Z, Higgs, leptons, and quarks. The photons and gluons act as mediators of electromagnetic and strong interactions, respectively. There are eight different types of gluons. The W^+, W^-, and Z particles mediate the weak interactions. All of these particles, that is, photons, gluons, W^+, W^-, Z, and Higgs are bosons. In contrast, leptons and quarks are fermions. These have spin $1/2$ in units of $\hbar$. We next list some basic properties of these particles:

1. Leptons: Leptons are particles such as electrons, muons, and neutrinos that do not participate in strong interactions. The known leptons include electron, muon, τ-lepton, electron neutrino, muon neutrino, and τ neutrino, denoted by

$$e^- \, , \, \mu^- \, , \, \tau^- \, , \, \nu_e \, , \, \nu_\mu \, , \, \nu_\tau \, ,$$

   respectively. We assign lepton number 1 to all of these particles. The corresponding anti-particles, positron, anti-muon, anti-τ-lepton, electron anti-neutrino, muon anti-neutrino, and τ anti-neutrino are denoted by

$$e^+ \, , \, \mu^+ \, , \, \tau^+ \, , \, \bar\nu_e \, , \, \bar\nu_\mu \, , \, \bar\nu_\tau \, ,$$

   respectively. These are assigned lepton number -1. The total lepton number is conserved in all reactions.

2. Quarks: In analogy with leptons, there are six different types or flavors of quarks, called up, down, strange, charm, bottom, and top. These are denoted by

$$u \, , \, d \, , \, s \, , \, c \, , \, b \, , \, t \, .$$

   Here we have listed them in increasing order of their masses. The electric charge of these quarks are $2/3, -1/3, -1/3, 2/3, -1/3, 2/3$, respectively, in units of the charge of proton. We assign baryon number $1/3$ to each of these quarks. The corresponding antiparticles are denoted by

$$\bar u \, , \, \bar d \, , \, \bar s \, , \, \bar c \, , \, \bar b \, , \, \bar t \, .$$

Each of these are assigned baryon number $-1/3$. Just as the lepton number, the baryon number is also conserved in all reactions.

A surprising fact about quarks is that they are never observed as free particles. For example, we can see a track of an electron inside particle detectors. Despite extensive experimental effort, it has not been possible to observe a similar track of a quark. Quarks always remain confined inside hadrons. By definition, hadrons are particles that participate in strong interactions. These are classified into baryons, which are fermions, and mesons, which are bosons. Baryons are bound states of three quarks, whereas mesons are composed of a quark and an anti-quark. Familiar examples of baryons are protons and neutrons. Because these are composed of three quarks, their baryon number is 1. Similarly, an anti-proton has baryon number equal to -1. All mesons are very short-lived particles and decay very quickly after being produced in a laboratory.

Baryon number conservation is responsible for the stability of protons and other atomic nuclei. This is because proton is the lightest baryon. Hence, by baryon number and energy conservation, it cannot decay. It can annihilate into two photons if it encounters an anti-proton. However there do not exist any anti-protons in most regions of the Universe today. On Earth they can only be produced in very high energy laboratories. In nuclear reactions in stellar cores also, anti-protons and anti-neutrons never appear since the energy is not sufficiently high to create these particles.

9.3 A Brief Introduction to Neutrinos

As we have already mentioned, there are three types of neutrinos: electron neutrino, muon neutrino, and τ neutrino, denoted by

$$\nu_e \, , \nu_\mu \, , \nu_\tau \, ,$$

respectively, as well as the anti-particles, electron anti-neutrino, muon anti-neutrino, and τ anti-neutrino, denoted by

$$\bar{\nu}_e \, , \bar{\nu}_\mu \, , \bar{\nu}_\tau \, ,$$

respectively. Neutrinos are electrically neutral, have spin $\hbar/2$, and have very small mass, $m_\nu < 1$ eV/c^2. They interact only through weak interactions. At energies on the order of 1 MeV, relevant to nuclear reactions, their interactions are extremely weak compared to other particles. Hence their mean free path through matter is very large, on the order of 10^{20} cm in the Earth or Sun.

This is roughly equal to 10^9 times the solar radius. Hence both the Earth and Sun are transparent to neutrinos. A beam of neutrinos can pass through the Earth or Sun without significant attenuation. Due to their large mean free path, they are able to emerge even from very dense astrophysical regions. For example, we are able to directly observe the neutrinos produced in the core of the Sun. In contrast, the photons are unable to escape from the core. The electromagnetic radiation we see originates near the surface of the Sun. The observed flux of neutrinos from the Sun, with energies on the order of 1 MeV, provides direct evidence that nuclear fusion reactions are the dominant source of solar energy.

9.4 PP Chain

As we have already mentioned the energy production in main sequence stars is dominated by the PP chain and the CNO cycle. These processes convert hydrogen to helium. Such reactions that release energy are called *exothermic*. In contrast, the reactions that consume energy are called *endothermic*. The PP chain consists of three separate chains, PP1, PP2, and PP3. The nuclear reactions corresponding to the PP1 chain are as follows:

$$
\begin{aligned}
{}^1\mathrm{H} + {}^1\mathrm{H} &\rightarrow {}^2\mathrm{H} + e^+ + \nu_e, \\
{}^1\mathrm{H} + {}^2\mathrm{H} &\rightarrow {}^3\mathrm{He} + \gamma, \\
{}^3\mathrm{He} + {}^3\mathrm{He} &\rightarrow {}^4\mathrm{He} + 2\,{}^1\mathrm{H}.
\end{aligned}
\tag{9.2}
$$

The symbol ${}^A\mathrm{X}$ denotes the nucleus X with mass number A. The positron produced in the first reaction combines with an electron in the medium to produce photons

$$
e^+ + e^- \rightarrow \gamma + \gamma.
\tag{9.3}
$$

The reactions produce energy in the form of kinetic energy of nuclei and photons that enhance the temperature of the medium. Some energy is also carried by the neutrino. This does not get converted into thermal energy because the neutrino escapes the Sun without any interactions.

The first reaction in Equation 9.2 involves both weak and strong interactions. The weak interactions convert one of the protons, $({}^1H)$, into a neutron. The proton and the neutron then undergo fusion to form a deuteron. Hence this process is very slow. Because this has the smallest reaction rate among all the processes in the PP1 chain, it determines the total rate of the PP1 chain. Interestingly, the slow rate of this process is responsible for the long lifetime of the Sun. The PP1 chain produces about 85% of the solar energy. The remaining 15% is produced by the PP2 chain and a very small fraction by the PP3 chain. The first two reactions are common to all three PP chains.

The subsequent PP2 chain reactions are

$$^3\text{He} + {}^4\text{He} \rightarrow {}^7\text{Be} + \gamma,\tag{9.4}$$

$$^7\text{Be} + e^- \rightarrow {}^7\text{Li} + \nu_e,\tag{9.5}$$

$$^7\text{Li} + {}^1\text{H} \rightarrow 2\,{}^4\text{He}.\tag{9.6}$$

The corresponding reactions for the PP3 chain are

$$^3\text{He} + {}^4\text{He} \rightarrow {}^7\text{Be} + \gamma,\tag{9.7}$$

$$^7\text{Be} + {}^1\text{H} \rightarrow {}^8\text{B} + \gamma,\tag{9.8}$$

$$^8\text{B} \rightarrow {}^8\text{Be} + e^+ + \nu_e,\tag{9.9}$$

$$^8\text{Be} \rightarrow 2\,{}^4\text{He}.\tag{9.10}$$

The PP chain dominates the solar energy production.

Besides the PP chain, the CNO cycle also contributes to the energy production of the main sequence stars. The PP chain dominates for stars whose mass is less than 1.3 solar mass. The CNO cycle dominates for higher mass stars. For such stars, the central temperature is higher. The CNO reaction rate increases much more rapidly with temperature in comparison to that of the PP chain. Hence it dominates energy production in stars of larger mass. The dominant CNO cycle consists of the following series of reactions:

$$^{12}\text{C} + {}^1\text{H} \rightarrow {}^{13}\text{N} + \gamma,\tag{9.11}$$

$$^{13}\text{N} \rightarrow {}^{13}\text{C} + e^+ + \nu_e,\tag{9.12}$$

$$^{13}\text{C} + {}^1\text{H} \rightarrow {}^{14}\text{N} + \gamma,\tag{9.13}$$

$$^{14}\text{N} + {}^1\text{H} \rightarrow {}^{15}\text{O} + \gamma,\tag{9.14}$$

$$^{15}\text{O} \rightarrow {}^{15}\text{N} + e^+ + \nu_e,\tag{9.15}$$

$$^{15}\text{N} + {}^1\text{H} \rightarrow {}^{12}\text{C} + {}^4\text{He},\tag{9.16}$$

where, besides helium, the reaction produces photons, electrons/positrons, and neutrinos. Notice that this set of reactions does not lead to any change in the abundance of carbon, nitrogen, and oxygen. These act as catalysts.

9.5   Nuclear Reaction Rate

In the main sequence phase, the energy of stars is provided by the conversion of hydrogen into helium. Inside the stellar cores, the temperature is so high that all the atoms are completely ionized. Let's consider the basic fusion reaction that converts two protons into a deuteron nucleus:

$$2\,{}^1\text{H} \rightarrow {}^2\text{H}\,.$$

In order to undergo fusion, the protons have to overcome the Coulomb barrier and come within a distance of 1 fermi of each other. At this distance, the nuclear attractive forces dominate over the Coulomb repulsion and fusion can take place. The electromagnetic Coulomb potential energy between two particles of charge Z_1e and Z_2e at distance r is given by

$$U(r) = \frac{Z_1 Z_2 e^2}{r} \,, \tag{9.17}$$

in CGS units. This is illustrated in Figure 9.1. For short distances, approximately less than 1 fm, the attractive nuclear potential dominates. The potential energy becomes negative in this region. The precise form of the nuclear potential is unknown theoretically. Here we have simply assumed that it is independent of r.

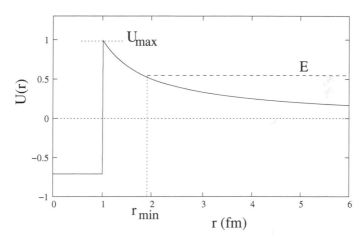

FIGURE 9.1: The nuclear and Coulomb potential experienced by two nuclei undergoing fusion is shown schematically by the solid curve. The repulsive Coulomb potential dominates for r greater than about 1 fm. At short distances, the attractive nuclear potential, which is assumed to be constant here, dominates. The dashed line represents the total energy of a particle that, classically, is unable to cross the Coulomb potential barrier.

Within the framework of classical mechanics, fusion is possible only if the energy of the particles is sufficient to cross the Coulomb barrier. Let us, for example, consider two nuclei, of masses m_1 and m_2 with velocities $\vec{v}_1$ and $\vec{v}_2$ and position vectors $\vec{r}_1$ and $\vec{r}_2$, respectively. Their total energy can be written as

$$E = K + U \,, \tag{9.18}$$

where K and U represent, respectively, their kinetic and potential energies. The kinetic energy is given by

$$K = \frac{1}{2}m_1 v_1^2 + m_2 v_2^2 \,. \tag{9.19}$$

We define the relative velocity $\vec{v}$ between the two particles as

$$\vec{v} = \vec{v}_1 - \vec{v}_2 . \tag{9.20}$$

Their center of mass velocity is given by

$$\vec{V}_c = \frac{m_1 \vec{v}_1 + m_2 \vec{v}_2}{m_1 + m_2} . \tag{9.21}$$

In terms of these, we can express K as

$$K = \frac{1}{2} \frac{m_1 + m_2}{2} V_c^2 + \frac{1}{2} \mu v^2 , \tag{9.22}$$

where

$$\mu = \frac{m_1 m_2}{(m_1 + m_2)} \tag{9.23}$$

is called the reduced mass. The first and second terms on the right-hand side of Equation 9.22 represent the kinetic energies of the center of mass motion and the relative motion, respectively. The center of mass motion is not of interest here. As we will explicitly see later, it does not play any role in the formula for reaction rate. Hence we will set $V_c = 0$ by going into the center of mass frame and focus here on the relative motion. The energy of the two nuclei system is then

$$E = \frac{1}{2} \mu v^2 + U(r) . \tag{9.24}$$

We can now treat this two-particle system as a single particle of mass μ at position $\vec{r} = \vec{r}_2 - \vec{r}_1$ and velocity $\vec{v}$.

For very large r, $U(r) \to 0$ and hence v is maximum. As the two particles approach one another, r decreases and the Coulomb potential starts to rise. Their relative speed decreases. Let us assume that the energy E is less than U_{max}, the maximum height of the potential, as shown in Figure 9.1. Classically, the two particles will then approach within a minimum seperation, r_{min}, at which point $v = 0$. This is indicated in Figure 9.1 by the point where E (dashed line) meets the potential energy curve. Beyond this time they will again start moving apart and hence will not be able to undergo fusion. Thus, classically, fusion is possible only if $E \geq U_{max}$. The energy of the protons follows the Maxwell–Boltzmann distribution. At any temperature T, there would be a few protons that have enough energy to overcome the Coulomb barrier and undergo fusion.

The process is properly treated within the framework of quantum mechanics. The phenomenon of quantum tunneling considerably enhances the probability for fusion to occur. In this case, fusion can occur even if the proton does not have enough energy to overcome the Coulomb barrier, that is, $E < U_{max}$. This can be understood by realizing that in quantum mechanics, the position of a particle is not well defined. The uncertainty in the position can be obtained by Heisenberg's uncertainty principle. The probability for a particle to be at position x is given by $\psi^*(x)\psi(x)$, where $\psi(x)$ is the wave function. The wave function does not become identically zero even in regions where the particle is prohibited by classical mechanics. These are regions it is unable to reach classically due to insufficient kinetic energy. However, because its wave function is nonzero, the probability of finding a particle, even in these regions, is nonzero. This leads to a much higher probability for fusion in comparison to what is predicted by classical mechanics.

We next obtain the formula for the reaction rate for this process. For this we require the velocity or energy distribution function of protons or nuclei. This is given by the Maxwell–Boltzmann distribution function (Section 8.9),

$$n(v_x, v_y, v_z)d^3v = N \left(\frac{m}{2\pi kT}\right)^{3/2} e^{-mv^2/2kT} d^3v, \qquad (9.25)$$

where N is the total number density of particles. We next need the probability for the fusion reaction to occur. This is expressed in terms of the cross section, $\sigma(E)$, discussed in Chapter 7. This can be visualized as the effective cross-sectional area of the target particle, as discussed earlier. If the incoming particle strikes within this area, then the reaction occurs. Otherwise the two particles do not scatter.

We have discussed some examples of cross section in Chapter 7. Here we consider another interesting example of the scattering of an electron on a charged target, which may be a proton, nucleus, or another electron. In this case, the interaction between the two particles is given by the Coulomb potential, which is a long-range potential. This means that it is not negligible even when the particles are very far apart, or when the impact parameter is very large. Within the framework of classical mechanics, the total scattering cross section can be deduced by solving Newton's equations of motion. In this case, one finds that the total cross section is infinite. A more appropriate quantity in this case is the differential cross section $d\sigma/d\cos\theta$, which is finite and measures the probability of scattering at an angle θ (for details see *Classical Mechanics* by H. Goldstein).

In quantum mechanics we cannot precisely specify both the position and momentum of the particle simultaneously. Hence it is meaningless to say that we have an incident particle of momentum $\vec{p}$ striking a target at impact parameter b as we did in Chapter 7 (see Figure 7.17). Here the cross section simply represents the probability that an incident particle of momentum $\vec{p}$ may scatter from a target. However, we can keep the same physical picture in mind to get an intuitive understanding the concept of cross section.

We next give a precise definition of cross section, valid both in classical and quantum mechanics. Consider a beam consisting of a large number of identical particles scattering from a target. The cross section between the beam and target particles is defined as

$$\sigma(v) = \frac{\text{number of reactions/target particle/time}}{\text{number of incident particles/area/time}}. \qquad (9.26)$$

The denominator in this equation is simply the beam flux. In order to measure the cross section between two particles, an experimentalist can set up one of these as beam particles, of known flux, and the other as target particles, at rest in the laboratory, of known density. The experimentalist would measure the number of reactions that occur over a given interval of time. Then the cross section can be extracted by using Equation 9.26. Alternatively, once we know the cross section for scattering between two particles, we can use this equation to determine the reaction rate in any experiment.

Before considering the reactions in the solar core, let's consider an idealized situation where the target particles are all at rest and the beam particles are all moving with the same velocity $\vec{v}$. This is applicable to many laboratory nuclear experiments, where, very often, one has a target at rest in the laboratory that undergoes collision with a uniform beam of particles. Even in the case of two uniform colliding beams, one can go into a frame in which one set of particles is at rest. The conditions in the solar core, however, differ from this as both colliding particles follow Maxwell-Boltzmann distribution. Hence they move in all directions and there is no frame in which the target particles are all at rest.

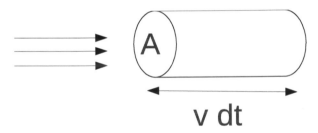

FIGURE 9.2: A beam of particles with velocity $\vec{v}$ incident normally on a cross-sectional area A. The number of particles crossing the area A in time dt is equal to the number of particles in length $v\,dt$ of the cylinder.

Let us denote the incident particle by 1 and the target by 2. We are interested in determining R_{12}, which is defined as the total number of reactions per unit time per unit volume. Then,

$$\text{number of incident particles/area/time} = N_1 v, \qquad (9.27)$$

where N_1 is the number density of incident particles. We can easily obtain this equation by considering a cylinder with its axis parallel to beam velocity, as shown in Figure 9.2. Let the cross sectional area of the cylinder be A. The number of particles crossing A in time dt is equal to the number of particles in length $v\,dt$ of the cylinder, which is equal to $N_1 v\,dt A$. Hence,

$$\text{number of incident particles/area/time} = \frac{N_1 v A dt}{A dt},$$

which gives Equation 9.27. Using Equation 9.26, we find that the reaction rate per unit volume, R_{12}, can be expressed as

$$R_{12} = N_1 N_2 \sigma(v) v \,, \tag{9.28}$$

where N_2 is the density of target particles. If the target is also in motion, then $\vec{v}$ is equal to the relative velocity between the beam and target particles.

We have so far obtained the reaction rate for a very special case, target particles at rest and beam particles moving in a particular direction. In the stellar interior, particles are moving randomly in all possible directions. Their velocities are specified by the Maxwell-Boltzmann distribution. There is no frame in which we can reduce the problem to the simple case discussed above. We derive the general result below. Here we first give the final answer. In the general case, the reaction rate per unit volume, R_{12}, can be expressed as

$$R_{12} = \lambda_{12} N_1 N_2 \,, \tag{9.29}$$

where $\lambda_{12} = \langle \sigma(v) v \rangle$ is the average of the cross section, $\sigma(v)$ times v over the velocity distribution of the particles in the medium. The precise expression is given below. We next give a detailed derivation of this formula.

9.5.1 Nuclear Reaction Rate: Derivation

We first assume that the target particles are at rest. The beam particles have a velocity distribution $n_1(v_x, v_y, v_z)$. The rate equation can be obtained by applying the arguments for the case of uniform beam to beam particles in an infinitesimal velocity interval, d^3v, and then integrating over the velocity. We obtain

$$R_{12} = \int \sigma(v) v n_1(\vec{v}) N_2 d^3 v \,. \tag{9.30}$$

The distribution function is normalized such that

$$\int n_1(\vec{v}) d^3 v = N_1 \,,$$

the total number density of beam particles. All the quantities inside the integrand depend only on the magnitude of the velocity $\vec{v}$. Hence we can also express the rate as

$$R_{12} = N_1 N_2 < \sigma(v) v > \,, \tag{9.31}$$

where

$$< \sigma(v) v >= \int_0^\infty \sigma(v) v f(v) dv \,, \tag{9.32}$$

and $f(v)$ is a distribution function, normalized such that

$$\int_0^\infty f(v) dv = 1 \,.$$

In the case of fusion reactions inside the solar core, the distribution function of both the beam and target particles is described by the Maxwell-Boltzmann distribution function. Let us denote the beam and target velocities by $\vec{v}_1$ and $\vec{v}_2$ respectively and the velocity distribution of target particles by $n_2(v_{2x}, v_{2y}, v_{2z})$. Consider the target particles in an infinitesimal velocity interval d^3v_2. The number density of these particles is equal to $n_2 d^3 v_2$. The reaction rate over this infinitesimal interval is given by

$$dR_{12} = n_2(\vec{v}_2)d^3v_2 \int \sigma(v)vn_1(\vec{v}_1)d^3v_1 \,. \tag{9.33}$$

Here v denotes the magnitude of the relative velocity $v = |\vec{v}_1 - \vec{v}_2|$. Finally integrating over d^3v_2 we obtain the total reaction rate as

$$R_{12} = \int \sigma(v)vn_1(\vec{v}_1)n_2(\vec{v}_2)d^3v_1 d^3v_2 \,. \tag{9.34}$$

This gives us

$$R_{12} = N_1 N_2 \int \sigma(v)v \left(\frac{m_1}{2\pi kT}\right)^{3/2} \left(\frac{m_2}{2\pi kT}\right)^{3/2} \exp\left(-\frac{m_1 v_1^2}{2kT} - \frac{m_2 v_2^2}{2kT}\right) d^3v_1 d^3v_2 \,.$$

We next change variables from $\vec{v}_1$, $\vec{v}_2$ to center of mass velocity $\vec{V}_c$ and relative velocity $\vec{v}$. The relationship between the two is given in Equation 9.64. The Jacobian of this transformation is unity. Hence we find that

$$R_{12} = N_1 N_2 \frac{(m_1 m_2)^{3/2}}{(2\pi kT)^3} \int \sigma(v)v \exp\left(-\frac{(m_1 + m_2)V_c^2}{2kT} - \frac{\mu v^2}{2kT}\right) d^3V_c d^3v \,,$$

where μ is the reduced mass. Because the distributions are normalized,

$$\int d^3V_c \left(\frac{m_1 + m_2}{2\pi kT}\right)^{3/2} \exp\left(-\frac{(m_1 + m_2)V_c^2}{2kT}\right) = 1 \,,$$

we again obtain Equation 9.29 with,

$$< \sigma v > = \left(\frac{\mu}{2\pi kT}\right)^{3/2} \int v\sigma(v) \exp\left(-\frac{\mu v^2}{2kT}\right) 4\pi v^2 dv \,. \tag{9.35}$$

In this derivation we have assumed that the two colliding particles are distinct. In case these two particles are identical, then the factor $N_1 N_2$ must be replaced by $N^2/2$, where N is the number density of the particles. The factor of $1/2$ arises in order to avoid double counting. We can obtain this by counting the number of distinct pairs we can make for a system consisting of N particles, with $N \gg 1$.

9.5.2 Nuclear Cross Section

We next theoretically obtain a formula for the cross section of a fusion reaction between two nuclei. This is given by fundamental physics and involves several concepts. A detailed derivation is beyond the scope of this book and here we simply provide an outline. The cross section has dimensions of area. Hence we can express it as

$$\sigma(E) \propto \lambda^2 \, ,$$

where λ is some suitable length scale associated with the process and $E = \mu v^2/2$ is the kinetic energy, ignoring the center of mass motion. The relevant length scale here is the de Broglie wavelength $\lambda = h/p$, where $p = \mu v$. Hence we obtain

$$\sigma \propto \frac{h^2}{p^2} \propto \frac{1}{E} \, .$$

We next need to include the factor corresponding to the quantum mechanical probability for barrier penetration. This is proportional to $\exp(-2\pi^2 U/E)$, where U is the height of the Coulomb barrier (see Equation 9.17). If $U << E$, barrier penetration is exponentially suppressed. Using $r \approx \lambda$ and $E = \mu v^2/2$, we find

$$\frac{U}{E} = \frac{2Z_1 Z_2 e^2}{vh} \, . \tag{9.36}$$

Collecting all the factors, we can express σ as

$$\sigma(E) = \frac{S(E)}{E} e^{-b/\sqrt{E}} \, , \tag{9.37}$$

where

$$b = 2^{3/2}\pi^2 Z_1 Z_2 e^2 \sqrt{\mu}/h \, . \tag{9.38}$$

The overall proportionality constant and the interaction strength between the colliding nuclei, which depends on energy, have been absorbed in the factor $S(E)$. In most cases, it is a slowly varying function of energy. But in some special cases, such as resonant scattering, it shows strong energy dependence.

9.5.3 Estimating the Nuclear Reaction Rate

Using Equation 9.37 we can express the reaction rate (see Equations 9.29 and 9.35) as

$$R_{12} = \left(\frac{2}{kT}\right)^{3/2} \frac{N_1 N_2}{\sqrt{\mu\pi}} \int_0^\infty S(E) \exp\left(-\frac{b}{\sqrt{E}} - \frac{E}{kT}\right) dE \, . \tag{9.39}$$

Here we have used $E = \mu v^2/2$ and changed the variable of integration from v to E. We plot the function, $g(E) = \exp\left(-\frac{b}{\sqrt{E}} - \frac{E}{kT}\right)$, in Figure 9.3. Let us assume that $S(E) \approx 1$. In that case the integral is simply the area under the curve shown in Figure 9.3. The Boltzmann factor $e^{-E/kT}$ falls rapidly with energy, whereas the barrier penetration factor $e^{-b/\sqrt{E}}$ rises with energy. At low energies, the penetration factor dominates. Hence $g(E)$ rises with E. Beyond a certain energy, the Boltzmann factor dominates and leads to decay with E. The product of these two factors is very small for most values of E. It peaks at

$$E_0 = \left(\frac{bkT}{2}\right)^{2/3}. \tag{9.40}$$

This is called the Gamow peak. It is clear from Figure 9.3 that the dominant contribution to R_{12} is obtained from the region in the neighborhood of the Gamow peak. The contribution from the regions far removed from the peak is very small. Hence we can evaluate the integral in the reaction rate R_{12} approximately by expanding the exponent around $E = E_0$.

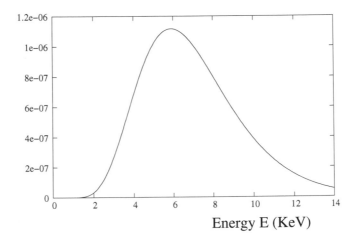

Energy E (KeV)

FIGURE 9.3: The factor $g(E) = \exp\left(-\frac{b}{\sqrt{E}} - \frac{E}{kT}\right)$ plotted as a function of energy, E. Here the temperature has been set equal to 1.5×10^7K. The Gamow peak occurs at energy $E_0 \approx 5.9$ KeV.

The integral is of the form

$$I = \int_0^\infty dE S(E) e^{-f(E)}, \tag{9.41}$$

where

$$f(E) = \frac{b}{\sqrt{E}} + \frac{E}{kT}. \tag{9.42}$$

Here $S(E)$ is assumed to be a slowly varying function of E, and $f(E)$ positive and much larger than unity. We expand $f(E)$ around the neighborhood of its extremal point,

$$\frac{df(E)}{dE}\bigg|_{E=E_0} = 0\,,$$

where E_0 is given by Equation 9.40. Hence we have

$$f(E) = f(E_0) + \frac{(E-E_0)^2}{2}\frac{d^2 f}{dE^2}(E_0) + \dots . \tag{9.43}$$

Keeping only the quadratic terms, the integral can be performed by setting $S(E) \approx S(E_0)$. We find

$$I = S(E_0)e^{-f(E_0)}\sqrt{\frac{2\pi}{f''(E_0)}}\,. \tag{9.44}$$

For the first reaction of the PP1 chain, $S(E_0) \approx 3.78 \times 10^{-22}$ KeV·barns.

There are several corrections to the above formula. Here we briefly discuss two:

(1) In the case of resonant scattering, the factor $S(E)$ also shows a strong dependence on energy. Hence the integrand in Equation 9.39 is considerably modified in this case.

(2) The medium is electrically neutral because besides protons and nuclei, it also contains electrons. This partially screens the nuclear target. Hence the repulsion due to Coulomb potential is somewhat reduced, leading to an increased barrier penetration rate. This enhances the reaction rate.

So far we have obtained the reaction rate for an individual process involving the collision of two particles designated 1 and 2. In practice we need to consider a chain consisting of several processes. The rate at which the final products of such a chain are produced is somewhat complicated because one has to solve differential equations involving the reaction rates of all the individual processes. Here we illustrate this calculation by considering the example of the PP1 chain in some detail.

Let's denote the number densities of hydrogen, deuteron, helium-3, and helium-4 nuclei by N_H, N_D, $N_{He(3)}$, and $N_{He(4)}$, respectively. The reaction rate for the pp collision, the first reaction in the PP1 chain, can be written as

$$R_1 = \lambda_1 \frac{N_H^2}{2}\,. \tag{9.45}$$

The rate of formation of ^{3}He through the second reaction in the PP1 chain can be expressed as,

$$R_2 = \lambda_2 N_H N_D\,. \tag{9.46}$$

The rate at which the concentration of deuteron changes with time is given by

$$\frac{dN_D}{dt} = R_1 - R_2 = \lambda_1 \frac{N_H^2}{2} - \lambda_2 N_H N_D\,. \tag{9.47}$$

The reaction strength λ_2 is much larger than λ_1. Let us assume that initially $N_D = 0$, which implies that $R_2 = 0$. As deuteron is produced in the pp fusion reaction, N_D increases and the reaction rate R_2 also starts increasing. At some stage, $R_1 = R_2$ and the deuteron concentration reaches an equilibrium value. Beyond this time, $dN_D/dt = 0$. This also gives us the equilibrium concentration of deuteron by the following formula,

$$\frac{N_D}{N_H} = \frac{\lambda_1}{2\lambda_2} . \tag{9.48}$$

Because $\lambda_2 \gg \lambda_1$, deuteron reaches its equilibrium concentration very quickly during the life cycle of a star.

The final reaction in the PP1 chain, which involves fusion of two ^{3}He is relatively slow in comparison to the second reaction. This is due to the larger Coulomb barrier. Hence it takes a much longer time for ^{3}He to reach its equilibrium value inside stellar cores. In fact, it may not be at its equilibrium concentration at all stages of the lifetime of a star, even when it is on the main sequence. The reaction rate for this process can be expressed as

$$R_3 = \lambda_3 \frac{N_{He(3)}^2}{2} . \tag{9.49}$$

The rate at which the ^{3}He concentration increases is given by

$$\frac{dN_{He(3)}}{dt} = R_2 - 2R_3 = \lambda_2 N_H N_D - 2\lambda_3 \frac{N_{He(3)}^2}{2} . \tag{9.50}$$

The extra factor of 2 multiplying R_{33} is due to the fact that two ^{3}He nuclei are consumed in each reaction. Here we will only consider the case when ^{3}He concentration has also reached its equilibrium value. In this case, $dN_{He(3)}/dt = 0$ and hence $R_2 = 2R_3$. This gives us the equilibrium concentration of ^{3}He,

$$\frac{N_{He(3)}}{N_H} = \left(\frac{\lambda_2}{2\lambda_3}\right)^{1/2} , \tag{9.51}$$

where we have used Equations 9.50 and 9.48. Finally we calculate the rate at which ^{4}He is generated. This is given by

$$\frac{dN_{He(4)}}{dt} = R_3 = \frac{R_2}{2} = \frac{R_1}{2} . \tag{9.52}$$

The rate R_1 can be evaluated analytically by making an expansion around the stationary point of $f(E)$, as explained earlier in this section. Collecting all the factors, we find

$$R_1 = \frac{N_H^2}{2} \left(\frac{2}{kT}\right)^{3/2} \frac{1}{\sqrt{\mu\pi}} S(E_0) e^{-f(E_0)} \sqrt{\frac{2\pi}{f''(E_0)}} . \tag{9.53}$$

9.6 Energy Released in Nuclear Reactions

We next determine the energy released in nuclear reactions taking the PP1 process as an example. The energy is released if the total mass of the final products is less than the mass of the initial particles. The total energy produced is given by

$$\Delta E = \Delta M c^2, \qquad (9.54)$$

where $\Delta M = M_i - M_f$, where M_i and M_f are the sum of masses of the initial and final state particles, respectively. In estimating the energy released, we should discount the energy carried away by the neutrinos produced because these simply escape from the star. Here the particles under consideration are bare nuclei with all their electrons stripped off. However, for the purpose of estimating ΔM, it is more useful to use the masses of the corresponding atoms. This is because the atomic masses are measured much more accurately in comparison to the nuclear masses. The atomic mass differs from the nuclear mass by the mass of the added electrons and the atomic binding energy. Let's ignore the atomic binding energies because these are very small compared to the typical energy released in nuclear reactions. Then by replacing nuclear masses by atomic masses, we would simply be adding the masses of equal number of electrons to both the initial and final masses. Hence we would only make a very tiny error, corresponding to the atomic binding energies, in making this replacement.

The atomic masses are given in atomic mass units (a.m.u.), which sets the mass of ^{12}C atom exactly equal to 12. In MeV units, it is given by

$$1 \text{ a.m.u.} = 931.478 \text{ MeV}/c^2.$$

The masses of different elements in a.m.u. are very close to their mass numbers A. Hence the atomic masses are often tabulated in the form of mass excess, M.E.(A,Z), defined as

$$\text{M.E.}(A, Z) = M(A, Z) - A,$$

where $M(A, Z)$ is the mass of the atom with mass number A and atomic number Z in atomic mass units. The tabulated values are often given by converting M.E.(A, Z) into energy in MeV by multiplication with the factor 931.478. The mass excess for neutron, ^{1}H, ^{2}H, ^{3}He, ^{4}He are 8.07132, 7.28897, 13.13572, 14.93120, 2.42491, respectively, in MeV units.

We next estimate the energy released in the PP1 chain. The first two reactions $^1\text{H} + {}^1\text{H} \rightarrow {}^2\text{H} + e^+ + \nu_e$ and $^1\text{H} + {}^2\text{H} \rightarrow {}^3\text{He} + \gamma$ essentially involve the conversion $3\,^1\text{H} \rightarrow {}^3\text{He} + e^+ + \nu_e + \gamma$. Hence the energy released is equal to $\Delta E_1 + \Delta E_2 = 3\text{M.E.}(1, 1) - \text{M.E.}(3, 2) = 3 \times 7.28897 - 14.93120 = 6.93571$ MeV. Here we have denoted ΔE_1 and ΔE_2 as the energy released in the first and second reaction, respectively. A relatively small fraction of this is carried away by neutrino. Here we will ignore this. The rate of energy generation in these two reactions is equal to $R_1 \Delta E_1 + R_2 \Delta E_2$. In equilibrium, $R_1 = R_2$. Hence the rate is simply equal to $R_1 \times 6.93571$ in units of MeV per unit volume per unit time. The energy released in the final reaction in the PP1 chain is equal to $2 \times 14.93120 - (2.42491 + 2 \times 7.28897) = 12.86$ MeV.

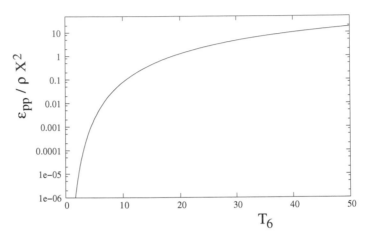

FIGURE 9.4: The rate of energy generation, ϵ_{PP1} in the PP1 chain, divided by ρX^2, as a function of $T_6 = T/10^6$K.

We define ϵ as the rate of energy generation in units of energy per unit mass per unit time. Hence $\rho\epsilon$ is the rate of energy generation in units of energy per unit volume per unit time. Here ρ is the density of the medium. Therefore, the total rate of energy generation in PP1 chain is

$$\rho\epsilon_{PP1} = R_1 \times 6.93571 + R_3 \times 12.86 = R_1 \times (13.37 \text{ MeV}), \qquad (9.55)$$

where ϵ_{PP1} is the energy released per gram per second in the PPI reaction. Using Equation 9.53 we can evaluate the energy generation rate as a function of temperature. It is useful to eliminate the number density of hydrogen nuclei, N_H, in terms of its mass fraction, X. The mass density of hydrogen is clearly ρN_H. Hence the number density of hydrogen is

$$N_H = \frac{\rho X}{m_H}, \qquad (9.56)$$

where m_H is the mass of hydrogen nucleus. Because $R_1 \propto N_H^2$, therefore $\rho\epsilon_{PP1} \propto N_H^2 \propto \rho X^2$. This implies that

$$\epsilon_{PP1} \propto \rho X^2. \qquad (9.57)$$

Using Equation 9.53 and collecting all the factors, we find

$$\epsilon_{PP1} = 2.36 \times 10^6 \rho X^2 T_6^{-2/3} \exp(-33.8 T_6^{-1/3}), \qquad (9.58)$$

in units of ergs/(g s). We plot this rate in Figure 9.4. Here $T_6 = T/(10^6 \text{K})$. If we assume that $PPII$ and $PPIII$ chains are subdominant, this gives an approximate formula for the energy generation rate, ϵ_{PP}, in the pp chain, assuming equilibrium concentration of ^{3}He.

In order to get a better idea about the dependence of ϵ_{PP1} on temperature T, it is useful to express it as a power of T in the neighborhood of the temperature of interest. Here we are interested in the temperature of the solar core, which is around 1.5×10^7K. Hence we express Equation 9.58 as

$$\epsilon_{PP} \approx \epsilon_{PP1} = \epsilon_{0,pp}\rho X^2 T_6^\alpha . \tag{9.59}$$

By making a Taylor expansion around $T = 1.5 \times 10^7$K of both Equations 9.58 and 9.59 and equating the first two terms, we can determine $\epsilon_{0,pp}$ and α. We find

$$\epsilon_{PP} \approx \epsilon_{0,pp}\rho X^2 T_6^{3.90} , \tag{9.60}$$

where $\epsilon_{0,pp} = 1.12 \times 10^{-5}$ erg cm^3/(g^2 s).

We next briefly discuss the CNO cycle, which also contributes to the energy production in the main sequence stars. In this case, the energy production rate in the neighborhood of $T = 1.5 \times 10^7$K can be expressed as

$$\epsilon_{CNO} \approx \epsilon_{0,CNO}\rho X X_{CNO} T_6^{\alpha_{CNO}} , \tag{9.61}$$

where $\alpha_{CNO} \approx 20$. This is much larger than 3.9, the exponent found in the case of ϵ_{PP}. Hence the rate of energy production rises very rapidly with temperature. The core temperature increases with the mass of the star. Hence for high mass stars, energy production is dominated by the CNO cycle. From stellar model calculations, we find that for stars with mass $M < 1.3 M_{Sun}$, the PP chain dominates, whereas the CNO cycle dominates for stars with mass $M > 1.3 M_{Sun}$.

During the main sequence phase, the energy production of a star is dominated by processes that convert hydrogen to helium. Hence the helium abundance in the core increases with time. The star evolves slowly during this phase. The star undergoes a major change when the hydrogen abundance in the core becomes negligible. At this point, the nuclear reaction rate becomes very small and is unable to balance the gravitational attraction. The ^{4}He nucleus cannot undergo significant fusion reactions at this temperature. The core shrinks rapidly, converting gravitational potential energy into thermal energy. Hence the temperature in the core increases and eventually it becomes hot enough to start helium fusion. The details of this process will be discussed later. Beyond this point the star settles into the red giant or supergiant phase.

The fusion process of helium nuclei also encounters a barrier, reminiscent of the pp fusion. In this case, the problem arises due to unusual stability, that is, high binding energy, of ^{4}He in comparison to other nuclei with $Z \le 5$. For example, the fusion of a helium nucleus with hydrogen to form lithium is highly suppressed in stellar cores. This reaction is endothermic and its rate would be appreciable only at much higher temperatures. Helium fusion could occur by the process

$$^4\text{He} + {}^4\text{He} \rightarrow {}^8\text{Be},$$

which converts two helium nuclei to a beryllium (^{8}Be) nucleus. However, this isotope of beryllium is unstable and quickly decays into two ^{4}He nuclei. The lifetime of ^{8}Be is approximately 10^{-16} seconds. In order to form heavier elements, ^{8}Be has to capture another ^{4}He nucleus within this time interval. The resulting nuclear reaction,

$$^4\text{He} + {}^8\text{Be} \rightarrow {}^{12}\text{C},$$

forms carbon. This process practically acts as the fusion of three helium nuclei and is called the triple α process. Here α stands for the ^{4}He nucleus. The rate for this process is suppressed and is controlled by the lifetime of ^{8}Be. Its rate is appreciable only at much higher temperatures, found in the cores of giant or supergiant stars. The energy production rate for the triple α process, in the vicinity of temperature $T = 10^8$K, can be expressed as

$$\epsilon_{3\alpha} \approx \epsilon_{0,3\alpha}\rho^2 Y^3 T_8^{40}, \tag{9.62}$$

where $T_8 = T/10^8$K and $\epsilon_{0,3\alpha} = 4.4\times10^{-8}$ erg cm^6/(g^3 s). This is discussed in detail in *Principles of Stellar Evolution and Nucleosynthesis* by D. D. Clayton.

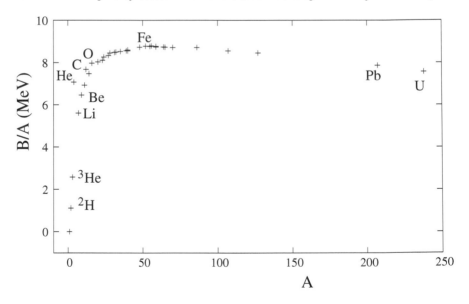

FIGURE 9.5: The binding energy B in MeV per nucleon for different nuclei. Here we show only the most abundant nucleus for each atomic number. The only exception is ^{2}H and ^{3}He, which are also shown.

The triple α process leads to the synthesis of carbon. At higher temperatures, fusion of heavier elements such as carbon, nitrogen, oxygen occurs and leads to synthesis of elements such ^{20}Ne, ^{23}Na, ^{24}Mg, ^{28}Si etc. We can obtain a qualitative understanding of the abundance of different nuclei by considering their binding energies B, defined as

$$B = Z \times m_p + (A - Z) \times m_n - M_A \,, \tag{9.63}$$

where m_p, m_n, and M_A are the masses of proton, neutron, and nucleus, respectively. The binding energy per nucleon (B/A) is plotted in Figure 9.5. We find that B/A increases up to iron, ^{56}Fe, and then starts to decrease. Hence fusion reactions that produce iron as their final product produce energy. In contrast, fusion reactions which produce nuclei heavier than Iron are endothermic, i.e. these consume energy. The latter reactions are not favored in equilibrium conditions inside stellar interiors. They occur by neutron capture processes at relatively high temperatures, often under special conditions such as supernova explosions. The abundance of such elements is relatively small. We also notice in Figure 9.5 that some nuclei such as ^{4}He, ^{12}C, and ^{16}O have relatively high binding energies in comparison to other nuclei of comparable atomic numbers. These turn out to be the most abundant elements, besides hydrogen, in the Universe.

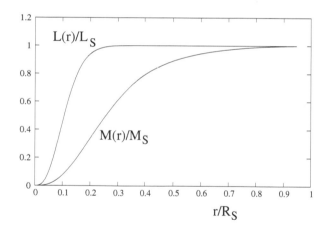

FIGURE 9.6: The mass and luminosity as a function of distance r from the center. On the x-axis we plot the distance r relative to the solar radius R_S. The mass and luminosity are shown relative to the total mass M_S and luminosity L_S of the Sun. (Data taken from Bahcall and Pinsonneault, *Phys. Rev. Lett.* 92, 121301 (2004).)

A star is formed by the collapse of a cloud of gas and dust in the interstellar medium. As the cloud collapses, its temperature increases. At some stage, the temperature in its core may rise high enough to start nuclear reactions that convert hydrogen to helium. At this point, the star becomes a main sequence star. The star leaves the main sequence once helium fusion starts. Because hydrogen fusion is a relatively slow process, the star spends maximum time on the main sequence. Hence we expect that the relative abundance of stars on the main sequence would be higher in comparison to other phases. This explains the main feature of the HR diagram, where it is found that most stars lie on the main sequence. Within the main sequence, the lifetime of a star decreases with an increase in its mass. The hydrogen fusion rate is higher in stars with larger mass due to their higher core temperatures. The lower limit on mass of a main sequence star is $0.08 M_{Sun}$. Stars with mass smaller than this do not have sufficiently high temperatures to start nuclear fusion. The upper limit is $90 M_{Sun}$. Stars with mass higher than this are unstable. Numerical model calculations are able to well reproduce the luminosity-temperature relationship of the main sequence stars.

9.7 Standard Solar Model

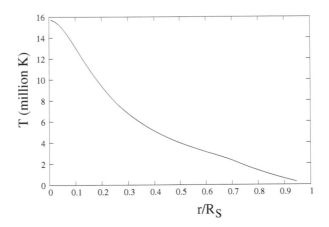

FIGURE 9.7: The temperature as a function of r/R_S. (Data taken from Bahcall and Pinsonneault, *Phys. Rev. Lett.* 92, 121301 (2004).)

The stellar model equations can be used to deduce the internal properties of stars. Here we provide details of the internal properties of the Sun obtained by a numerical solution of these equations. In Figure 9.6 we show the mass and luminosity of the Sun from the center to its surface. We find that the luminosity reaches close to the surface luminosity at $r \approx 0.2 R_S$, where R_S is the solar radius. This is expected because the nuclear reactions take place only in the core. In Figure 9.7 we show the temperature as a function of r. The temperature decreases from 16 million K at the center to about 0.6 million K at $r = 0.9R_S$. In Figure 9.8, we show how the composition changes as a function of the radius. Here we only show the variation of mass fractions of hydrogen, X, and helium, Y, with radius. As expected, we find that the fraction of helium is much higher in the thermonuclear core, about 0.65 near the center, and decreases rapidly to about 0.3 at $r = 0.2R_S$. At the surface, the helium mass fraction is about 0.25. The energy transport inside the Sun is dominated by radiation for $r < 0.7R_S$. At larger r, convection dominates.

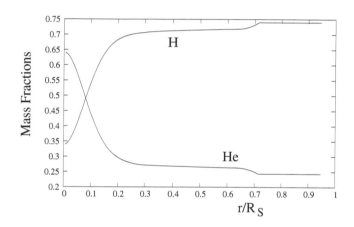

FIGURE 9.8: The mass fractions X and Y of hydrogen, ^{1}H, and helium, ^{4}He, respectively as a function of r/R_S. (Data taken from Bahcall and Pinsonneault, *Phys. Rev. Lett.* 92, 121301 (2004).)

Exercises

9.1 Which of the following reactions are allowed by the conservation laws:

$$e^- + {}^1\text{H} \rightarrow \gamma + \gamma \,,$$

$$n \rightarrow {}^1\text{H} + e^- + \nu_e \,,$$

$$^3\text{He} + {}^3\text{He} \rightarrow {}^4\text{He} + {}^2\text{H} \,.$$

If a reaction is not allowed, state which conservation law is violated. Note that here ^{1}H refers to hydrogen nucleus, which is simply a proton.

9.2 Let $\vec{v}_1$ and $\vec{v}_2$ be the velocities of two particles of masses, m_1 and m_2, respectively. Show that, in terms of the relative velocity $\vec{v}$ and center of mass velocity $\vec{V}_c$, these can be expressed as

$$\vec{v}_1 = \vec{V}_c + \frac{m_2}{m_1 + m_2}\vec{v}\,,$$
$$\vec{v}_2 = \vec{V}_c - \frac{m_1}{m_1 + m_2}\vec{v}\,. \tag{9.64}$$

9.3 Starting from the basic formula for the kinetic energy of two particles, Equation 9.19, derive Equation 9.22.

9.4 Express the distribution function, $f(v)$, defined by Equation 9.32, in terms of the distribution $n_1(v)$.

9.5 Verify that the Gamow peak position occurs at an energy given by Equation 9.40.

9.6 Make a Taylor expansion of ϵ_{PP1} given in Equations 9.58 and 9.59 around $T = 1.5 \times 10^7$K. Note that $T = 1.5 \times 10^7$K implies $T_6 = 15$. Keep only terms up to first order in $(T_6 - 15)$. Equate the zeroth and first-order terms of the two expansions and determine the parameters α and $\epsilon_{0,pp}$.

9.7 Verify the formula, Equation 9.58 for ϵ_{PP1}. Proceed as follows: (a) Determine the formula for R_1 by explicitly substituting the expressions for $f(E_0)$ and $f''(E_0)$ in Equation 9.53. (b) Express the formula for $\rho\epsilon_{PP1}$, Equation 9.55, in units of ergs cm^{-3} sec^{-1}. (c) Obtain the expression for ϵ_{PP1} by using the results of (a), (b) and Equation 9.56. (d) The value of $S(E_0)$ is given by, $S(E_0) \approx 3.78 \times 10^{-22}$ KeV·barns Convert this into cgs units. Note that 1 barn is equal to 10^{-24} cm^2. (e) Numerically compute the exponent and the coefficient in order to obtain Equation 9.58.

9.8 (a) Make a rough estimate of the total number of neutrinos emitted by the Sun per second. Assume that only the PP1 chain contributes. By comparing the solar luminosity with the energy emitted per reaction, determine the total number of fusion reactions per second. Hence determine the neutrino emission rate. Compute also the solar neutrino flux at Earth. (b) The neutrino-proton cross section for typical solar neutrino energies is roughly 10^{-41} cm^2. Assume that an experiment to detect solar neutrinos uses a detector of mass 10,000 Kg. Make a rough estimate of the number of protons in this detector. Using the solar neutrino flux, determine the number of events, that is, neutrino-proton reactions, per second for this detector.

9.9 Check that the Jacobian of the transformation between $(\vec{v}_1, \vec{v}_2)$ and $(\vec{v}, \vec{V}_c)$ given in Equation 9.64 is unity. You only need to consider any particular component, x, y, or z of these velocity vectors and show that the corresponding Jacobian is unity.

Chapter 10

Star Formation and Stellar Evolution

The interstellar medium is filled with gas and dust. In some regions the density of gas and dust is much higher than the mean value in the medium. Stars are formed due to the collapse of such clouds or regions of high density. The temperature of these clouds is normally quite small and hence they do not emit any visible radiation. However, they are often illuminated by light due to stars in their neighborhood. Such illuminated clouds are called nebula. Due to their high density, they cause considerable extinction of star light. Hence the density of stars appears much reduced in the direction of these clouds.

The gas in the interstellar medium is predominantly hydrogen and helium. Hydrogen may be in the atomic, ionized, or molecular state and contributes roughly 70% by mass. Helium is predominantly in the atomic state and approximately 28% by mass. The mean gas density is about 10% of the total density of the Milky Way. This amounts to about 1 atom/cm^3. Dust is composed of complex molecules formed out of atoms such as carbon, oxygen, nitrogen, silicon, and hydrogen. Its mean density is about 0.1% of the total density of the galaxy. On an average, their number density is about 10^{-13} cm^{-3}. The size of the dust particles is typically less than 1 μm.

We find several different types of clouds in the interstellar medium. We observe diffuse hydrogen clouds, where hydrogen is found in its atomic state. The temperature in these clouds ranges from 30K to 80K. The number density lies in the range 100 to 800 cm^{-3}, and their total mass is on the order of 1 to 100 M_{Sun}. Due to their low temperature, hydrogen is found in its ground state. We also observe molecular hydrogen clouds. These are regions of high density of molecular hydrogen and dust. Molecular hydrogen would normally be broken down by star light. However, these clouds are shielded by regions of high-density atomic hydrogen and dust. These cause considerable extinction of star light and hence allow molecular hydrogen to exist. We observe many enormous complexes of dust and gas, called Giant Molecular Clouds (GMCs). These have temperature, $T \approx 20$K, mean density $n \approx 100$ to 300 cm^{-3}, mass $M \approx 10^6 \, M_{Sun}$, and size about 50 pc. The cores of these clouds have much higher densities, $n \approx 10^7$ to 10^9 cm^{-3}, and slightly higher temperatures $T \approx 100$ to 200K. Their mass and size are on the order of 10 to 1000 M_{Sun} and 0.05 to 1 pc, respectively. These are the sites of active star formation. Many such clouds are found in the Milky Way. They exist predominantly in the spiral arms.

10.1   Early Stage of Star Formation

During the initial stage of star formation, a compact object called a protostar is formed. This is the stage before the nuclear reactions start. A molecular cloud will collapse if it is gravitationally unstable. We can formulate the condition for collapse by considering the virial theorem, Equation 5.23. In equilibrium, $2K + U = 0$, where K is the total kinetic energy and U the potential energy. If $2K > |U|$, then the gas pressure dominates over the gravitational attraction and the cloud expands. On the other hand, if $2K < |U|$, then the cloud is unstable to collapse. Normally a cloud remains in equilibrium until some trigger, such as a nearby supernova explosion, makes it unstable to collapse. Consider a uniform spherical cloud of radius R and density ρ_0. Its mass $M = 4\pi R^3 \rho_0 / 3$ and its gravitational potential energy (see Exercise 8.5)

$$U = -\frac{3}{5} \frac{GM^2}{R} . \tag{10.1}$$

We assume that the cloud has the same temperature T throughout, and hence its total kinetic energy K is given by

$$K = \frac{3}{2} NkT , \tag{10.2}$$

where $N = M/(\mu m_H)$ is the total number of particles and μ the mean molecular weight. The condition for collapse, $2K < |U|$, implies that

$$3NkT < \frac{3}{5} \frac{GM^2}{R} .$$

We, therefore, find the condition

$$M > M_J \tag{10.3}$$

for collapse, where M_J is the Jean's mass,

$$M_J = \left(\frac{3}{4\pi}\right)^{1/2} \left(\frac{5k}{Gm_H}\right)^{3/2} \left(\frac{T^3}{\mu^3 \rho_0}\right)^{1/2}. \qquad (10.4)$$

This formula should be treated as an order of magnitude estimate due to our assumption that the density of the cloud is uniform.

We next apply the Jean's criteria, Equation 10.3, to the cores of the Giant Molecular Clouds. Using $T = 100K$, $n \sim 10^8$ cm^{-3} or $\rho_0 \sim 10^{-16}$ g/cm^3, we find $M_J \sim 10\ M_{Sun}$. Hence, the observed masses of these cores, of order 10 to 1,000 M_{Sun}, satisfy the criteria for collapse. In contrast, one can check that the atomic hydrogen clouds do not satisfy the Jean's criteria.

These clouds usually stay in equilibrium until a trigger starts their collapse. The trigger may be some disturbance such as a supernova explosion in their neighborhood which sends out a shock wave into the interstellar medium compressing the gas with which it comes in contact. During the initial stages, the cloud is nearly in free-fall under gravitational attraction, with the pressure gradient being almost negligible. The cloud has sufficiently low density that heat is efficiently radiated out. Hence its temperature does not change substantially during collapse and remains the same throughout the medium. The collapse at this stage is isothermal. One can solve Newton's equation of motion to determine the typical time scale of collapse. One finds the time scale for free-fall collapse,

$$t_{ff} \approx \left(\frac{1}{G\rho_0}\right)^{1/2}. \qquad (10.5)$$

Using $\rho_0 \sim 10^{-16}$ g/cm^3 we find $t_{ff} \sim 5,000$ years. Hence we find that this time scale is very short compared to both the thermal time scale t_{th} and the time scale for evolution of a star driven by nuclear fusion.

As the cloud contracts, its density increases. In time, its core becomes sufficiently dense so that the heat generated due to the release of gravitational potential energy does not get radiated away efficiently. The temperature of the core starts to increase. The pressure gradient can no longer be neglected and the core contracts at a slower rate. The surrounding medium, however, is still in free-fall. The central object is called a protostar. For a cloud of mass M_{Sun}, this region would have a size on the order of a few AU. The temperature of this object is such that it radiates at infrared frequencies. The energy transport occurs dominantly by convection. This object is surrounded by dense molecular clouds. Hence observational evidence for such an object is the existence of small IR sources embedded within molecular clouds. Their detection is difficult because these objects are very short lived due to the smallness of the free-fall time scale.

The free-fall phase eventually stops and the cloud achieves hydrostatic equilibrium. At this point, this object is located on the HR diagram at the extreme right-hand side on a line called the Hayashi track. An object cannot be in hydrostatic equilibrium toward the right of this track. Toward the left, hydrostatic equilibrium can be maintained. At the Hayashi track, the star is fully convective. After reaching this track, the star contracts quasi-statically on the thermal time scale with the pressure gradient balancing the gravitational attraction. A star of mass equal to M_{Sun} spends about 10^7 years and a star of mass 15 M_{Sun} spends about 60,000 years in this phase. Once the core temperature gets sufficiently high, fusion reactions start. Initially these give subdominant contribution to the total energy released by the star. The star reaches main sequence once the fusion reactions start dominating the total energy production.

10.1.1   Fragmentation

Stars often form in clusters. This is due to the fact that as the cloud collapses, it undergoes fragmentation, wherein the different fragments collapse independently to form stars. The fragmentation occurs because the original cloud is not exactly homogeneous. It has some regions of higher density. The Jean's mass M_J is proportional to $1/\rho_0^{1/2}$. Hence, as the collapse proceeds, some of these regions may themselves satisfy the Jean's criteria due to their higher density. During the initial stages of collapse, the temperature of a cloud does not change. However, its density increases. Hence in time, some dense regions might themselves become unstable to collapse and start contracting independently of the original cloud, leading to fragmentation. This region may undergo further fragmentation due to its inhomogeneous density distribution. The process of fragmentation would continue until our assumption that the collapse is isothermal breaks down. At this point, the cloud becomes very dense and the gravitational potential energy released cannot be efficiently transported out. The temperature starts to rise and the process of fragmentation stops.

   The large O, B type stars form first. When these stars reach the main sequence, their UV radiation raises the temperature of the medium. This inhibits further star formation. Furthermore, their radiation pressure drives out the outer layer of the cloud. This may disperse the remaining cloud. Due to the increase in temperature and reduction in cloud size, stars that were gravitationally bound earlier may no longer remain bound and may start to drift apart. Young clusters dominated by O, B stars are called OB associations. The newborn stars are revealed once the medium becomes sufficiently transparent. Planets form in the medium surrounding a star during its early stages of formation due to condensation of gas and dust. Observational examples of pre-main sequence stars are the T Tauri stars. These objects represent a transition from the infrared sources that are surrounded by dense clouds of dust and gas and the main sequence stars. The young stars also emit jets. When these jets meet the interstellar medium, they produce bright star-like nebula called Herbig–Haro objects.

10.2 Evolution on the Main Sequence

A star reaches the main sequence once nuclear fusion starts dominating its energy production. During this stage the energy of the star is generated by fusion processes that convert hydrogen into helium. A star spends maximum time in this phase. This explains why most of the stars lie along the main sequence on the HR diagram. For the lower main sequence, which corresponds to mass $M < 1.3 M_{Sun}$, the PP chain dominates. Here the energy transport in the core is dominated by radiation. Convection dominates in the outer regions. As the star evolves on the main sequence, hydrogen in the core gets depleted. Eventually the medium near the center gets converted entirely into helium and fusion reactions stop. However, hydrogen fusion continues in the surrounding shell, as shown in Figure 10.1.

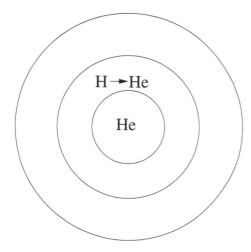

FIGURE 10.1: The fate of a low mass star toward the end of the main sequence. The core becomes predominantly helium, and hydrogen fusion continues in the shell surrounding the core.

The lower limit of the mass of a star is $0.08\ M_{Sun}$. If the mass of an object is below this limit, it does not get hot enough to start nuclear fusion. Stars with masses between $0.08\ M_{Sun}$ and $0.26\ M_{Sun}$ are fully convective on the main sequence. Hence the medium gets completely mixed up and all the hydrogen is eventually converted into helium. Such a star does not get hot enough to start helium fusion. The fusion reactions stop beyond this stage, the star begins to contract and becomes a white dwarf at the end of its life cycle. A white dwarf is a compact star with very high density. A white dwarf with mass equal to M_{Sun} has a radius close to the Earth's radius. Its pressure is dominated by degenerate free electron gas. This means that the density of free electrons is so high that all their quantum mechanical energy levels are filled up. The pressure exerted by this gas is sufficient to balance the gravitational attraction. We discuss this further in Section 10.3.

The energy production of stars in the lower main sequence, $M < 1.3\,M_{Sun}$, is dominated by the PP chain. Excluding the very low mass stars, $M < 0.26\,M_{Sun}$, these stars have a radiative core. Convective transport dominates in the outer regions. In the upper main sequence, $M > 1.3\,M_{Sun}$, the CNO cycle dominates. In this case, the rate of energy production has a strong dependence on temperature. Because the temperature decreases with distance, r, from the center, the energy production rate also has a strong dependence on r. Hence the outward flux of energy is very large, and can be maintained only by convection. The core of upper main sequence stars is, therefore, convective.

For stars with mass, $M > 0.26\,M_{Sun}$, the fusion reactions stop inside the core after all the hydrogen is converted to helium. The hydrogen burning continues inside an envelope surrounding the core. The helium core continues to expand as more and more helium is produced. Once the helium core is sufficiently large, it becomes unstable to collapse. The mean molecular weight μ of helium is larger than hydrogen. Because the Jean's mass decreases with an increase in μ, $M_J \propto 1/\mu^{3/2}$, a helium core of smaller mass in comparison to the hydrogen core may be unstable to collapse. As the core contracts, its temperature increases. Eventually the temperature becomes sufficiently large to start helium fusion. The extra energy produced pushes out the envelope surrounding the core. The star approaches the giant or supergiant phase, depending on its mass. Its luminosity increases but its surface temperature goes down.

The evolution of a star beyond the main sequence depends strongly on whether or not the core acquires electron degeneracy. Hence we next study this phenomenon in more detail.

10.3   Degenerate Free Electron Gas

We consider a gas of free electrons in volume V. As explained in section 6.5, electrons are fermions and hence only one electron can occupy any particular quantum state. At low temperatures their distribution function is similar to the $T = 0$ distribution function plotted in Figure 6.8. Hence, in this case, all the available quantum states or energy levels up to some maximum energy ϵ_f, called the Fermi energy, are filled. Such a gas is called degenerate. The Fermi energy ϵ_f is equal to the chemical potential μ, defined in Equation 6.11, in this case. The maximum momentum p_f corresponding to this maximum energy is called the Fermi momentum. The number of states in the momentum interval

p to $p + dp$ was derived in Chapter 6 (see Equation 6.32). It is given by

$$dN = 2\frac{4\pi V p^2 dp}{h^3},\tag{10.6}$$

where the factor 2 is due to the two spin states of an electron. The total number of states with $p < p_f$ is

$$N = \int_0^{p_f} 2\frac{4\pi V p^2 dp}{h^3} = \frac{8\pi V p_f^3}{3h^3}.\tag{10.7}$$

At high temperature, the states are sparsely filled. There are several states available with momentum in the neighborhood of a given state. If we heat this system, then the electrons can freely make transitions into states with momentum (and energy) close to the electron's momentum. This is not possible for a degenerate gas. Electrons cannot freely jump from one state to another. This also implies that electrons are not able to undergo scatterings since a scattered electron would necessarily have a different momentum in comparison to the incident electron. Hence it must occupy a different quantum state. This makes the degenerate matter highly conducting.

We next obtain the equation of state for the degenerate gas. Since all the states up to the Fermi momentum are filled up, the number density (n_e) of free electrons in such a system is given by $n_e = N/V$. The Fermi momentum is given by

$$p_f^3 = \frac{3n_e h^3}{8\pi}.\tag{10.8}$$

For a nonrelativistic gas, the kinetic energy of an electron with momentum p is

$$K = \frac{p^2}{2m_e},\tag{10.9}$$

where m_e is the electron mass. Hence the total energy of the gas is

$$E = \int K dN = \int_0^{p_f} \frac{4\pi V p^4 dp}{m_e h^3} = \frac{4\pi V p_f^5}{5m_e h^3}.\tag{10.10}$$

The pressure P can be computed by using Eq. 8.43. We obtain

$$P = \frac{2}{3}\frac{E}{V} = \frac{1}{20}\left(\frac{3}{\pi}\right)^{2/3}\frac{h^2}{m_e}n_e^{5/3},\tag{10.11}$$

where we have used Equation 10.8. This is the equation of state of a degenerate free electron gas. It is valid as long as the temperature of the electron gas is sufficiently small so that we can assume that it is degenerate. In a white dwarf, the gravitational attraction is balanced by the pressure of a degenerate electron gas, given by this equation.

10.4 Evolution beyond the Main Sequence

A star of mass less than 0.26 M_{Sun} becomes a white dwarf at the end of the main sequence phase. If the mass of a star is greater than 0.26 M_{Sun}, helium burning starts in the core. The star becomes a red giant or a supergiant. If the mass of a star lies between 0.26 M_{Sun} and 2 M_{Sun}, the helium fusion starts explosively. For higher mass stars, it starts non-explosively. This explosion is called a helium flash.

The explosion occurs because the fusion starts when the core is degenerate. In this situation the gravitational attraction is balanced by the pressure of the degenerate electron gas. The nuclear fusion reactions generate heat, leading to an increase in temperature. However the pressure does not change much as long as the temperature is sufficiently small so that degeneracy is maintained to a good approximation. Hence, in contrast to the non-degenerate matter, the core is unable to expand with increasing temperature. Meanwhile the nuclei get heated up. The reaction rate of helium fusion, that is, the triple α process, is very sensitive to temperature. Hence an increase in temperature leads to a rapid increase in the energy production rate. This leads to a further increase in temperature and triggers a runaway nuclear fusion reaction. Another important aspect of the phenomenon is that the conductivity of a degenerate electron gas is very high. Hence the temperature in this medium is uniform and the fusion process starts in the entire core almost at the same time. At sufficiently high temperature, the degeneracy is lifted and the core expands explosively. This is called a helium flash. The outer envelope is pushed out and the star settles into the red giant phase. Despite the explosion in the core, it does not lead to an overall disruption of the star.

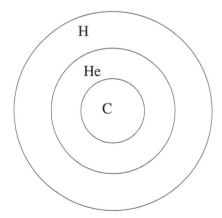

FIGURE 10.2: The fate of a star with mass less than $15 M_{Sun}$ during its late stages in the giant phase. The core becomes predominantly carbon with helium and hydrogen fusion occurring in shells surrounding the core.

If the mass of a star is less than $3M_{Sun}$, the core never gets hot enough to burn carbon. As carbon accumulates in the core, helium and hydrogen fusion takes place in the surrounding shells, as shown in Figure 10.2. Once the carbon core gets big enough, it becomes unstable to collapse. The collapse produces energy, which expels the outer envelope into the interstellar medium, forming, what is called a planetary nebula. We point out that, despite the name, this has nothing to do with planets. The core of this system forms a white dwarf.

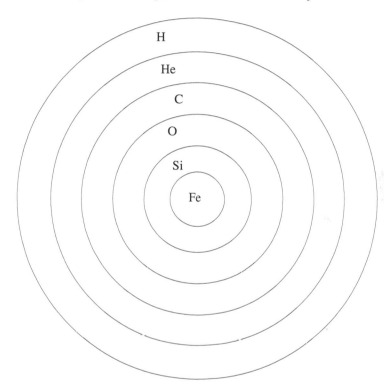

FIGURE 10.3: The fate of a star with mass greater than $15M_{Sun}$ during its late stages in the giant phase. The core becomes predominantly iron with elements of lower atomic weight burning in shells surrounding the core.

In the case of stars of mass larger than $3M_{Sun}$, the nuclear reactions proceed beyond helium fusion. The temperature in the core gets sufficiently high to burn carbon and heavier elements. If the mass lies in the range $3M_{Sun} < M < 15M_{Sun}$, the core eventually becomes degenerate. Its mass is dominated by carbon at the lower end of this mass range and by oxygen at the upper end. Hence, in analogy with a helium flash, the fusion of carbon (or oxygen) starts explosively. This is a very intense process and probably destroys the entire star in a supernova explosion.

If the mass of a star is greater than $15M_{Sun}$, the carbon fusion in the core starts when it is nondegenerate. Hence the process is nonexplosive. In this case, all the allowed fusion reactions take place, eventually forming iron in the core. During its final stages, the star has an iron core, with silicon, oxygen, carbon, helium, and hydrogen fusion occurring in shells, as shown in Figure 10.3. Fusion reactions cannot occur in equilibrium in the iron core. As its mass becomes sufficiently large, it becomes unstable to collapse. During the collapse, the gravitational energy released breaks down iron into lighter nuclei, which further break down, eventually forming protons and neutrons. This collapse occurs very rapidly, within a fraction of a second. The outer layer explodes as a supernova. The core collapse eventually stops when it reaches neutron degeneracy. In this case, the free neutron gas becomes degenerate. The star becomes a neutron star. If the mass of the star is sufficiently large, even the neutron degeneracy pressure is not enough to stop the collapse. The star becomes a black hole.

The upper limit on the mass of a white dwarf is

$$M_{wd} < 1.44 M_{Sun}. \tag{10.12}$$

This is called the Chandrasekhar limit. If the mass of a star is larger than this, then the electron degeneracy pressure is not enough to stop its collapse. The star becomes a neutron star. Similarly there is a limiting mass of the neutron star. Beyond this mass, the star becomes a black hole.

10.5   Population I and II Stars

The Big Bang model of the Universe proposes that the Universe originated at some time and has been expanding since then. In very early times, it was very hot and filled with plasma composed of photons, electrons, protons, neutrons, and neutrinos. Due to very high temperatures, neither nuclei nor atoms could exist at that time. As it cooled, a few light nuclei, predominantly helium, formed. As it cooled further, hydrogen and helium atoms formed. Eventually, galaxies and galaxy clusters formed. Hence the first stars in the Universe formed out of material that was predominantly hydrogen and helium. The percentage of other elements was negligible. These stars are called population II stars. These are the oldest stars in the Universe. Within the Milky Way, they are found in globular clusters, which populate the galactic halo. These clusters probably formed at the same time as the formation of the Milky Way. The surface composition of population II stars is predominantly hydrogen and helium. In the interior, of course, heavier elements are present due to fusion.

As the Universe evolves, some of the population II stars end their life in a supernova explosion, which replenishes the interstellar medium with metals, that is, elements with atomic number $Z > 2$. Hence the next and subsequent generation of stars is formed in a medium that is much richer in metals. These are called population I stars. For example, the Sun is a population I star. These have a much larger abundance of metals on their surface in comparison with population II stars. They are found predominantly in the disks of spiral galaxies, in particular the spiral arms, and are generally younger, brighter, and hotter in comparison with population II stars. Because a higher mass star has a smaller lifetime, we expect that population II stars, in general, have lower masses in comparison with population I stars. The higher mass population II stars would have reached the end of their life cycle by the current time.

10.6 White Dwarfs

White dwarfs are compact stars formed at the end of the life cycle of a star of mass less than 3 solar masses. Recall that if the mass of a star lies between $0.08 M_{Sun}$ and $0.26 M_{Sun}$, then it never gets sufficiently hot to start helium fusion. Such stars become white dwarfs after exhausting their hydrogen fuel. These stars are predominantly composed of helium. In the case of higher mass stars, $0.26 M_{Sun} < M < 3 M_{Sun}$, fusion processes convert helium to carbon in the core. The core never gets hot enough to burn carbon and starts contracting once it becomes sufficiently large. The outer layers of the star form a planetary nebula while the core becomes a white dwarf. In this case, the white dwarf consists predominantly of carbon. If the mass of a star lies between $3 M_{Sun}$ and $15 M_{Sun}$, then the carbon or oxygen burning in the core starts explosively. In this case, the star may be completely destroyed in a supernova explosion. However if such a star loses a sufficient amount of mass during its life cycle, then it is possible that it may leave behind a remnant white dwarf composed predominantly of carbon and oxygen.

White dwarfs are supported by the electron degeneracy pressure, given by Equation 10.11, and hence are very dense. The radius of a white dwarf of mass equal to 1 solar mass is of the order of the Earth's radius. Hence a white dwarf is about a million times denser than the Sun and about 300,000 times denser than Earth. There is an upper limit on the mass of a white dwarf, which is given by Equation 10.12. No white dwarf has been observed that violates this limit.

By using the equation of hydrostatic equilibrium, we find that the mass of a white dwarf is inversely proportional to its volume,

$$M_{wd} \propto \frac{1}{V_{wd}}. \tag{10.13}$$

Hence in contrast to the main sequence and giant stars, a white dwarf of larger mass has a smaller radius. A white dwarf has no source of energy. It has no nuclear reactions. It also does not undergo contraction and hence does not release gravitational potential energy. Initially its temperature is relatively high. It slowly cools off by radiating away its thermal energy. Its luminosity decreases and it slowly fades away.

White dwarfs have very low luminosity compared to the main sequence stars. This is due to their very small size. Hence they are difficult to observe. An example of a white dwarf star is Sirius B, which is a binary partner of the bright star Sirius A. Another interesting example of a white dwarf is shown in Figure 10.4.

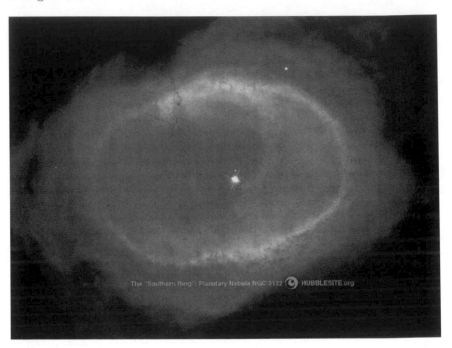

FIGURE 10.4: The Southern Ring Nebula is a planetary nebula illuminated by star light. The white dwarf associated with the planetary nebula is the faint star next to the bright star at the center of the nebula. (Image courtesy of NASA.)

10.7   Neutron Star

If the mass of a compact star is larger than the Chandrasekhar limit, then the electron degeneracy pressure is unable to stop its collapse. The star continues to collapse until the neutron gas in the star becomes degenerate. The gravitational attraction is balanced by the neutron degeneracy pressure. The stable object thus formed is called a neutron star. The formation of a neutron star is always accompanied by a supernova explosion. The medium of a neutron star consists predominantly of neutrons. An important process that leads to high density of neutrons is the URCA process. Consider a nucleus X with atomic number Z and atomic mass A. Under normal circumstances it will undergo the following processes:

$$
\begin{aligned}
{}^{Z}X + e^- &\rightarrow {}^{Z-1}Y + \nu_e, \\
{}^{Z-1}Y &\rightarrow {}^{Z}X + e^- + \bar{\nu}_e.
\end{aligned}
\tag{10.14}
$$

However, if the free electrons are degenerate, then the second process is suppressed. In a degenerate gas, all states up to the Fermi energy are filled. Hence the final state electron has no available energy level, thus leading to suppression. The first reaction in Equation 10.14 leads to a reduction in the atomic number of a nucleus. Hence the ratio of neutrons to protons in the final state nuclei becomes larger. Once this ratio becomes sufficiently large, the binding energy of nuclei starts to decrease. At this stage, neutrons start leaking out of the nuclei. These processes slowly convert all nuclei into neutrons. Eventually, the medium becomes very rich in neutrons, with only a small fraction of electrons and protons.

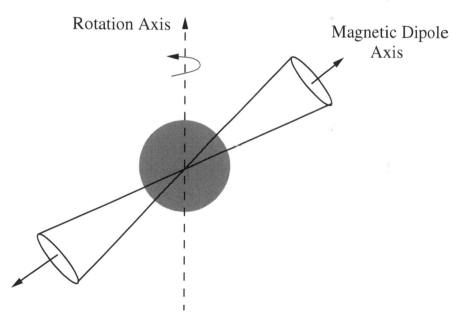

FIGURE 10.5: An illustration of a pulsar. The magnetic field axis is aligned at an angle to the rotation axis. Relativistic charged particles escape the pulsar spiraling along the magnetic field, emitting synchrotron radiation. The radiation is emitted in a cone centered on the magnetic field axis. The cone sweeps across the sky as the pulsar rotates. A detector receives radiation only for short time intervals when the emitted radiation is directed toward it.

Neutron stars are very compact objects. A 1 solar mass neutron star has a radius on the order of 10 Km. Its density is of order 10^{18} Kg/m^3. Observationally, neutron stars are identified with pulsars. The radiation from a pulsar is received in periodic pulses. The period of these pulses is observed to lie between a millisecond and a second. The period is found to be very precise after averaging over a series of pulses. Pulsars emit radiation predominantly in radio waves and x-rays. Their radiation at visible frequencies is observed to be very small.

The pulsed radiation is caused by rotation of the pulsar. Let's assume that the large-scale magnetic field of a pulsar is aligned at an angle to its rotation axis, as shown in Figure 10.5. As the pulsar rotates, it drags the magnetic field with itself. Relativistic charged particles emerge from pulsars spiraling along the magnetic field lines. As these particles accelerate in the magnetic field, they emit synchrotron radiation in a cone. This radiation appears as a pulse to an observer in its path. The period of the pulse is equal to the period of rotation of the pulsar.

10.8   Black Holes

If the compact star has sufficiently high mass, then even the neutron degeneracy pressure is unable to stop its collapse. The star eventually becomes a black hole. A black hole is a singularity in space-time and can be described only within the framework of the general theory of relativity. Let's assume a spherically symmetric mass distribution centered at the origin O. Let its total mass M be localized over a very small radius r. An object very far away from O will experience the gravitational force due to this object, roughly in accordance with what is implied by the Newton's law of gravitation. Hence it will not notice anything extraordinary. However, an object that falls within a radius

$$R = \frac{2GM}{c^2} \approx 2.95 \left[\frac{M}{M_{Sun}} \right] \text{Km} \tag{10.15}$$

will notice something unusual. It will be pulled toward this object and will not be able to escape its gravitational pull, no matter how hard it tries. This surface, beyond which nothing can escape, is called the event horizon. The radius R is called the Schwarzschild radius. We emphasize that nothing, not even light, can escape once it has entered the event horizon. As indicated in Equation 10.15, for a solar mass object, the Schwarzschild radius is about 2.95 Km.

FIGURE 10.6: An illustration of an accretion disk in a binary star system. Mass flows from the tear-shaped star at left to the one at right, where it forms an accretion disk. This process is discussed in more detail in Chapter 13. (Image courtesy of ST ScI, NASA.)

An object acts as a black hole only if its mass is concentrated in a radius less than the Schwarzschild radius. If an observer is very far away from a black hole, then it will not experience anything unusual. The gravitational attraction of a black hole at large distances is similar to that of any spherically symmetric object and can be described reliably by Newton's law of gravitation. However, at small distances, one has to use the general theory of relativity to properly describe a black hole. The nature of space-time changes completely in the vicinity of a black hole. Nevertheless, we get some understanding of the physics of black holes even by using Newtonian mechanics. Let's assume a spherical symmetric object of mass M. By the special theory of relativity, the maximum speed that any particle can attain is the velocity of light. Let's compute the radius at which the escape velocity becomes equal to the speed of light. This is found to be equal to the Schwarzschild radius. Hence we again get the result that nothing can escape from the Schwarzschild radius because the escape velocity at this surface approaches the maximum speed attainable by any particle. We hasten to add that this calculation should be taken with caution, as here we have used Newtonian mechanics, which do not have any concept of limiting velocity. This concept arises only when we use the special theory of relativity.

A black hole does not emit any radiation. Hence it is not possible to observe an isolated black hole. We can only observe it indirectly if it is gravitationally bound to other objects. For example, we can deduce the presence of a black hole in a binary system through the radiation emitted by its accretion disk. If a black hole is part of a binary star system, then it captures or accretes material from its binary partner. The matter falling on the black hole forms a disk-like structure called the accretion disk. This is shown schematically in Figure 10.6. As more matter falls on the black hole, it passes through the accretion disk and dissipates energy. This heats up the accretion disk, which then emits radiation at x-ray frequencies.

10.9   Supernova

At the end stage of its evolution, a star with sufficiently high mass explodes as a supernova. The luminosity of the star increases many fold, by as much as 100 million times within a few days. After reaching its peak value, the luminosity declines slowly over a long time interval. The outer shell is expelled into interstellar space at a very high velocity. The velocity may exceed the sound speed in the medium and hence this process produces a shock wave. The resulting remnant of the explosion is visible as a nebula.

Supernovas are classified into two types, Type I and II, depending on the shape of their light curves, that is, the variation in their intensity with time, and the observed spectral lines. Supernova type II arise due to the explosion of an isolated star whose mass is much larger than the mass of the Sun. The supernova type I arises due to collapse of a star in a binary system. In this case the mass of the star may be comparable to the solar mass.

Theoretically, we expect that there should be roughly one supernova explosion every 20 years in the Milky Way. However, it is not possible to observe all of these events because most of them occur close to the galactic center and are hidden from our view by clouds of gas and dust. The last major supernova observed at Earth occurred in 1987 and is called SN1987A. It occurred in a small, nearby galaxy called the Large Magellanic Cloud, LMC. It was a supernova of type II. The brightest recorded supernova event so far occurred in 1006 and is called SN1006. The exploding star was observable during the day for some time. The star appeared three times the size of Venus and had an intensity comparable to the Moon. Another interesting supernova explosion was SN1054. This was also observable during day time. It is possible to identify the remnant of this supernova as the Crab Nebula.

In the case of an isolated star of mass greater than 15 M_{Sun}, the supernova explosion arises due to the collapse of the iron core. The collapse occurs very rapidly, within a few seconds. This process releases a lot of energy in the form of radiation and neutrinos. The neutrinos immediately escape from the dense medium surrounding the core. Hence the first observable signal of a supernova explosion is a burst of neutrinos. The radiation produced during this process actually gets trapped and is released later, after a few hours. The outer envelope is thrown into interstellar space at velocities exceeding the sound velocity in the medium, producing shock waves. An image of the supernova remnant, such as the Crab Nebula, shows filamentary structure, which is evidence for these shock waves. There is also direct evidence for expansion of this nebula, that is, it has been possible to observe the increase in its size with time. The supernova shock waves are also believed to be sites where charged particles, observable on Earth as cosmic rays, are accelerated to ultra-high energies.

Exercises

10.1 The Jean's criteria can also be formulated in terms of the radius of a cloud. Consider a uniform cloud of density ρ_0, temperature T, and mean molecular weight μ. Show that it is unstable to collapse if,

$$R > R_J,$$ \hfill (10.16)

where R_J is the Jean's radius,

$$R_J = \left(\frac{15kT}{4\pi G m_H \mu \rho_0} \right)^{1/2}.$$ \hfill (10.17)

10.2 Verify the formula in Equation 10.5 for the time scale of free-fall collapse. Assume a cloud of initial density ρ_0 and radius R. Consider a test particle at its surface and determine its acceleration. Find the time it takes to reach the center, assuming that it is initially at rest and its acceleration constant. The latter assumption is justified because we are only interested in an order of magnitude estimate. Ignore numerical factors of order unity in the final formula.

10.3 Determine the Jean's mass for a Giant Molecular Cloud that has $T = 100K$ and $n = 10^8$ cm^{-3}. Repeat this exercise for an atomic hydrogen cloud that has $T = 30K$ and $n = 800$ cm^{-3}. Verify that this does not satisfy the criteria for collapse.

10.4 The Crab nebula is located at a distance of 2.0 ± 0.5 Kpc and has a mean angular diameter of roughly 5 arc minutes. Its angular diameter is increasing at the rate of approximately 1.6 arc seconds every 10 years. Determine the diameter of the nebula and the rate at which it is increasing.

10.5 Using Newtonian mechanics, determine the escape velocity of a particle located at the Schwarzschild radius. Verify that it is equal to the speed of light.

10.6 Verify Equation 10.13, which states that the white dwarf mass is inversely proportional to its volume. Start by using Equation 8.7. Let P be the mean pressure over the entire star. Hence argue that

$$\left|\frac{\Delta P}{\Delta r}\right| \approx \frac{P}{R} \approx \frac{GMM}{R^2 V}, \tag{10.18}$$

where we have set ρ approximately equal to the mean density M/V. Next, notice that in Equation 10.11, $n_e = n_p$, where n_p is the number density of protons. The mass of a star is predominantly provided by protons and neutrons. Hence up to a factor of order unity, $M \sim n_p V m_p$. This implies that $n_e \propto M/V$. Use Equation 10.11 and the above equation for P/R to establish Equation 10.13.

10.7 The Chandrasekhar limit on the mass of a white dwarf arises because the speed of the electrons becomes close to the speed of light once the mass of a white dwarf approaches this limit. You can verify this by using the rough estimate of pressure given in Equation 10.18. Eliminate the volume, V, in terms of the mass by using $M = \rho V$ and $\rho \approx 2 m_p n_e$. The formula for ρ follows because $n_e = n_p$, and it is reasonable to assume that $n_p = n_n$. Here n_p and n_n are the number densities of protons and neutrons, respectively. Next, eliminate P in terms of n_e using Equation 10.11. Finally compute the momentum and velocity of an electron using Equation 10.8. Compare with the speed of light for M equal to the limiting value.

Chapter 11

The Sun

The Sun is an average, middle-aged, main sequence star belonging to the spectral class G2. Like all other stars, it is composed of hot gas consisting predominantly of hydrogen and helium. It does not have any solid surface. The observed surface simply corresponds to the diffuse spherical region from where we receive the solar radiation. Due to its vicinity, we know a lot more about the Sun in comparison with all other stars. We have detailed observations in a wide range of frequencies, such as radio, infrared, visible, ultraviolet, and x-rays. Furthermore, we are also able to study the pressure (or sound) waves propagating in the solar interior. These produce oscillations in the solar surface that lead to Doppler shifts in the absorption spectral lines in the solar radiation. By measuring these Doppler shifts, astronomers can deduce the nature of the pressure waves in the solar interior. This phenomenon is similar to the seismic waves in Earth's interior and hence its study is called helioseismology. These observations provide a wealth of information about the Sun's interior. Besides these, we can also study neutrinos from Sun, which originate in the solar core due to nuclear reactions. These neutrinos provide direct information about the fusion reactions occurring in the core.

The Sun rotates about its axis and has a magnetic field. It does not, however, rotate as a rigid body. The speed of rotation is higher near the equator in comparison to the poles. Helioseismology has revealed that the angular speed also depends on the distance from the center. The latitude dependence of angular speed has a profound effect on the solar magnetic field, which varies over a 22-year cycle. This phenomenon is also correlated with the variation of Sun spots.

The surface temperature of the Sun is approximately 5,778K. It is a population I star, that is, it has a rich metal content, and was formed about 4.6 billion years ago. The solar spectrum, as seen on top of the atmosphere, is shown in Figure 6.3. The flux is dominant at visible frequencies. The solar flux density received at Earth is called the solar constant. It is equal to 1,360 W/m^2. The interior properties of the Sun can be deduced theoretically by solving the stellar model equations. The results of this study, called the Standard Solar Model, are discussed in Section 9.7.

11.1 Solar Atmosphere

The density of the Sun becomes very small beyond a certain radius. The region beyond this radius may be called the atmosphere, although, in contrast to Earth, there is no sharp boundary that marks the base of the atmosphere. The solar atmosphere is broadly divided into the photosphere, chromosphere, and corona, as shown in Figure 11.1. The photosphere is the base of the atmosphere. The upper most layer, that is, the corona extends to very large distances in the inter-stellar space.

11.1.1 Photosphere

The visible surface of the Sun lies in the photosphere. The Sun is opaque below this region. As we move outward in the photosphere, the opacity slowly decreases and we eventually reach the region of free emission. The attenuation of light, emitted from a point at radius r, is proportional to $\exp(-\tau)$, where

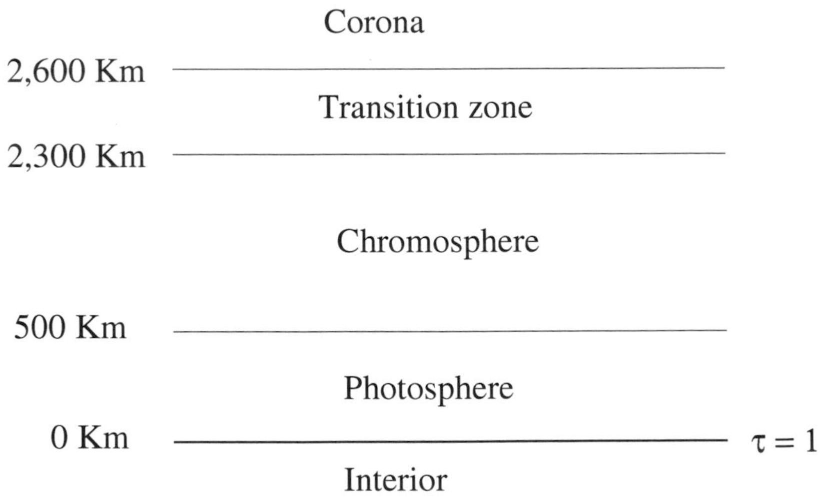

FIGURE 11.1: An illustration of the different layers in the solar atmosphere, from the photosphere to the corona.

τ is the optical depth, given by

$$\tau = \int_{r}^{\infty} \bar{\kappa}(r')\rho(r')dr' \,, \tag{11.1}$$

where $\bar{\kappa}$ is the Rosseland mean opacity and ρ is the density. Here we have assumed that the radiation propagates radially outward and the observer is located at very large distance, which may be taken to be infinity. We receive visible radiation from the photosphere, which is a narrow layer of thickness roughly 500 Km. Below the photosphere, $\tau \gg 1$, leading to very strong attenuation. The optical depth τ at the base of the photosphere is approximately equal to 1.

Near the surface, the dominant contribution to continuum opacity of the Sun comes from the H$^-$ ions, discussed in Section 8.6. These ions have ionization potential of 0.754 eV and can absorb photons at visible and higher frequencies.

The gas density in the photosphere varies roughly in the range 10^{-7} g/cm^3 to 10^{-9} g/cm^3 from the lower to upper boundary. For comparison, the density of air near the Earth's surface is of order 10^{-3} g/cm^3. Hence the photosphere is very sparse, with density almost 100,000 times smaller compared to air.

The temperature decreases with radius in the photosphere. At the base, the temperature is 6,500K. At the upper boundary the temperature is 4,400K. The effective temperature at the visible surface is 5,778K. The photosphere is cooler in comparison to the lower layers. Most of the absorption lines we observe in the solar spectrum are produced in the photosphere, due to its lower temperature. If we view the Sun radially, we get emission integrated over all the layers. However, if we view it tangentially, along the limb, as shown in Figure 11.2, we get emission only from the cooler regions in the photosphere. Hence near the surface, the Sun appears darker in comparison to the body. This phenomenon is called *limb darkening*.

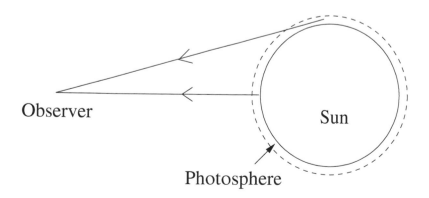

FIGURE 11.2: The photosphere of the Sun appears darker if viewed tangentially because it is cooler in comparison to the interior.

The photosphere appears granulated, as shown in Figure 11.3. The granulation arises due to convection cells formed by rising and falling air below the photosphere. The bright regions correspond to hot rising gas, which originates deep inside the Sun. As the gas rises, it cools and sinks in regions that appear darker. The difference in temperature in these two regions is found to be a few hundred degrees. These cells are continuously changing, typically lasting for a fraction of an hour, and have spatial extent of roughly 700 Km. The speed at which the gas rises near the center of a cell is measured by Doppler effect to be approximately 1 Km/sec. We also observe supergranules with diameters of about 30,000 Km. These change at a slower rate in comparison to granules, typically lasting for a few days.

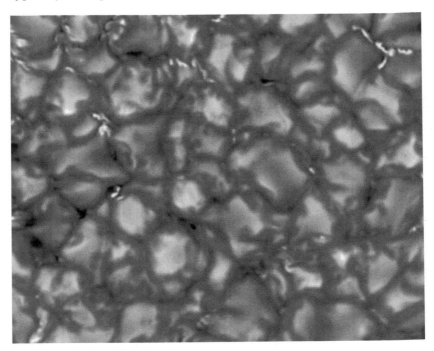

FIGURE 11.3: A picture of the solar surface showing its granulated structure. This image was observed by Vasco Henriques using the Swedish 1 m Solar Telescope (SST) at the Institute for Solar Physics, Stockholm, Sweden.

11.1.2 Chromosphere

The layer above the photosphere is called the chromosphere. Here the gas density is very low, smaller than the photosphere by a factor of 10^4. Furthermore, the intensity of radiation produced in this region is also very small. However, the temperature increases with radius in the chromosphere. It increases from 4,400K at the base to about 25,000K at the top. Due to high temperature, many atoms are found in the ionized state in the chromosphere. Furthermore, in some regions of the chromosphere, a significant fraction of the neutral atoms, such as hydrogen, are found in excited states. Hence several absorption and emission lines of atoms as well as ions, such as HeII, FeII, SiII, CrII, and CaII, are seen in this region. In the visible spectrum we observe only absorption lines because the emission lines are overwhelmed by the continuum emission from the photosphere. But at infrared and ultraviolet frequencies emission lines are also seen. Emission lines at visible frequencies may by seen during a total eclipse when most of the continuum emission is blocked out. This is called the flash spectrum. For example, one sees the Balmer H_α line, which gives the chromosphere a reddish appearance. The chromosphere may also be seen using a narrow filter that transmits only a narrow range of frequencies centered at this line.

The high temperature in the chromosphere means that particles in this region have very high velocities, given by the relationship, Equation 8.69, derived using the kinetic theory of gases. Due to low density, however, the energy per unit volume is very small. It is smaller than the energy density in the photosphere by a factor of more than 10^3. In contrast to the photosphere, the temperature in the chromosphere cannot be interpreted as the blackbody temperature. The fact that temperature increases with distance from the center is a puzzle, which is not satisfactorily resolved so far. We discuss this in more detail below.

11.1.3 Corona

Above the chromosphere we have the solar corona. The temperature rises very rapidly in the transition region, reaching roughly a million degrees in the corona, where it continues to increase gradually. The density in corona is very small, on the order of 10^{-16} g/cm^3 in the lower corona. This is roughly a factor of 10^4 smaller than in the chromosphere. Due to its very high temperature, the corona contains highly ionized atoms. Most of the light elements, such as H, He, C, N, and O, are completely ionized. Even heavier elements such as calcium and iron have most of their electrons removed. Hence one observes absorption and emission lines from highly ionized atoms from the corona. Many of these ions were unfamiliar to early astronomers. Hence they were not able to properly interpret this spectrum.

At visible wavelengths, the corona is not visible because its emission is very low compared to the photosphere. However, during total eclipse, the corona is observable at visible frequencies. An image of the corona during total solar eclipse is shown in Figure 11.4.

The corona has much higher emission in radio and x-rays. It emits radio waves by the process of free-free emission and synchrotron radiation. The x-ray emission is due to the atomic transitions which can produce high-frequency emission lines at such extreme temperatures. An x-ray image of the Sun, shown in Figure 11.5, directly reveals the corona. The image shows very uneven emission. We also see regions where the emission is extremely small. These are called coronal holes. These are regions of open magnetic field lines. Here the field lines from the Sun extend large distances out into the interstellar medium. Highly energetic charged particles, predominantly electrons and protons, which originate inside the Sun, are able to move freely parallel to the magnetic field lines. They escape freely from coronal holes and, hence, these regions have low plasma densities and temperatures. In contrast, charged particles get trapped in regions of closed magnetic field, producing high densities and temperatures. Hence the intensity of x-rays is very high in these regions.

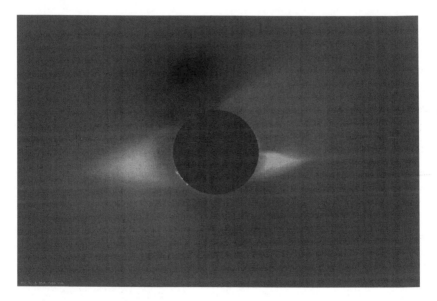

FIGURE 11.4: An image of the Sun during total eclipse reveals the corona (Image courtesy of High Altitude Observatory/NCAR/UCAR.)

The flux of energetic charged particles from the Sun is called solar wind. It is emitted continuously in all directions and extends to very large distances into the solar system. The speed of the solar wind ranges from 300 Km/s to 800 Km/s, with the higher speeds being produced by the coronal holes. The solar wind has a considerable effect on the Earth's magnetic field. Close to the surface of the Earth the magnetic field is approximately dipolar. However at higher altitudes it gets deformed with open field lines extending to large distances. The region above the dipolar magnetic field is called the magnetosphere. An illustration of how the solar wind forms the Earth's magnetosphere is shown in Figure 11.6. The charged particles in the solar wind become trapped in the Earth's magnetic field and produce the spectacular phenomenon of Aurora Borealis.

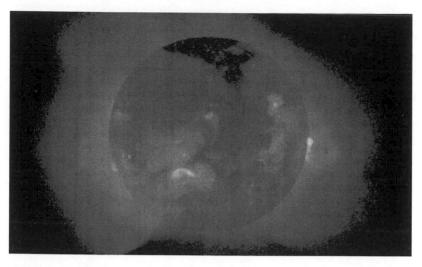

FIGURE 11.5: An image of the Sun at x-ray frequencies taken at the Yohkoh
Solar Observatory. (Yohkoh image is courtesy of the NASA-supported Yohkoh
Legacy data Archive at Montana State University.)

A surprising feature of the corona and chromosphere is the increase in
temperature with distance, r, from the center of the Sun. The temperature in
the corona rises to more than a million degrees. Because heat flows radially
outward from the Sun, it is more reasonable to expect a decrease in temperature with r. The reason for this increase is still not well understood. As in
the case of the chromosphere, we may use Equation 8.69 in order to relate
the temperature in the corona to the mean speed of particles. The energy per
unit volume decreases with r even in corona and chromosphere due to the
rapid decrease in gas density in these regions. In particular, the kinetic energy
density, that is, kinetic energy per unit volume, in the corona is two orders
of magnitude smaller than the kinetic energy density in the chromosphere.
Hence it is only the temperature and not the energy density that shows an
anomalous increase with r. It is also important to note that the solar corona is
far from thermal equilibrium. Hence the distribution of particle velocities does
not follow the Maxwell–Boltzmann distribution, and the concept of temperature is not applicable in the usual sense. In other words, we cannot reliably
use Equation 8.69 in order to relate temperature to the mean value of the
square of velocity, $< v^2 >$. Hence, rather than introducing the concept of
temperature, it might be better to directly consider the variable, $< v^2 >$. It
is really this variable that shows the anomalous increase as we move outward
from the photosphere.

There are many proposals to explain this behavior. Here we briefly explain
one of these mechanisms that involves heating by magnetic field reconnections.
The solar atmosphere has a complex structure of magnetic field flux tubes,
which extend to large distances beyond the photosphere. These flux tubes

get twisted and entangled due to changes in the solar atmosphere. Eventually they get realigned, releasing magnetic energy into the solar atmosphere that may accelerate particles to very high energies, thus leading to a large value of $< v^2 >$.

11.2 Dynamo Mechanism for Magnetic Field Enhancement

All astrophysical objects, the Sun, Earth, Milky Way, have a significant magnetic field. How is this field generated and maintained? There is no known mechanism by which these objects can generate a magnetic field by themselves. However, let us assume that initially there exists a small magnetic field, called the seed field. This seed field can be enhanced and maintained by the rotational and convective motion of these objects. This is called the

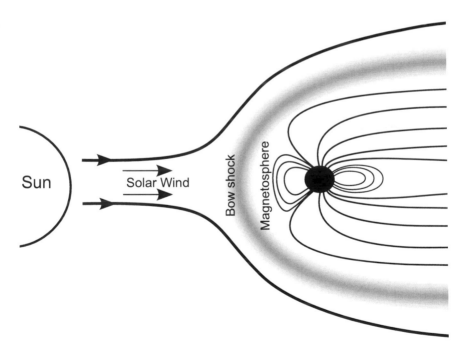

FIGURE 11.6: An illustration of the interaction of solar wind with the Earth's magnetic field. The solar wind is compressed and deflected by the Earth's magnetic field at the bow shock. Below the bow shock is the magnetosphere. Finally close to the surface of the Earth the field shows a dipolar behavior.

dynamo mechanism. One requires conducting fluid and sufficient mechanical energy for this purpose. The mechanical energy of rotation and convection is converted to magnetic energy. If there is no motion, the field will decay slowly with time. Recall that, except for ferromagnetic materials, none of the other materials can remain magnetized after the removal of the external magnetic field. Because astrophysical objects are not ferromagnetic, it is reasonable that without motion, their field will decay.

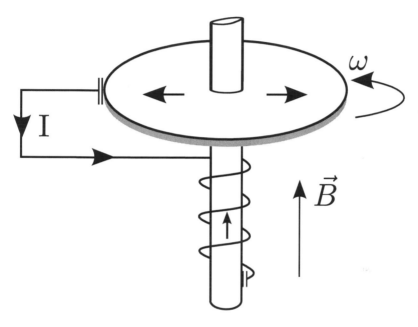

FIGURE 11.7: The dynamo mechanism for the enhancement of a magnetic field. The disc rotates with an angular speed ω in a background magnetic field $\vec{B}$. This process generates a current I which further enhances the magnetic field.

Let us now assume that there exists a small seed magnetic field. If the conditions are right, this field can be enhanced by mechanical motion. We explain this by taking a simple example. Consider a conducting disk, rotating counterclockwise, as shown in Figure 11.7. The disk is mounted on a conducting cylindrical rod at its center. There exists a background magnetic field pointing upward. A coil is connected between the outer edge of the disk and the rod by brushes. The coil remains stationary as the disk rotates. As the disk rotates, an electrical current flows through the disk and coil and back through the rod. Due to this current, an additional magnetic field is generated. The coil is oriented such that the magnetic field generated points in the same direction as the background magnetic field. Hence the overall strength of the field is enhanced. The enhancement of the field leads to increased current, which further enhances the field. The energy of the increased field and current is provided by the mechanical force that rotates the disk. The field saturates once the current becomes so large that the energy is dissipated at the same rate as it is supplied.

In the Sun and other astrophysical objects, a similar mechanism is operative, powered by rotation and convection. These objects have plasma that can sustain electric currents. Given the right circumstances, the magnetic field can be enhanced by the dynamo mechanism. The detailed mechanism in such objects is complicated and not well understood so far. As we will see, the solar magnetic field shows a periodic variation that is closely tied to the variation in the number of sunspots. This periodic variation is believed to be one of the important clues to the dynamo operational inside the Sun.

11.3   Sunspots and the Solar Cycle

A fascinating phenomenon observed on the solar surface is the appearance of sunspots. These are regions where the intensity is much smaller than the mean intensity on the surface. The intensity is smaller due to lower temperature. The surface temperature at a sunspot is typically about 4,000K in contrast to the mean surface temperature of 5,770K. The number of sunspots varies periodically with time with a period of about 11 years. The variation in sunspot number with time is shown in the lower plot of Figure 11.3. The plot shows the percentage of area of the visible solar hemisphere covered by sunspots, averaged over a day.

The average latitude at which the sunspots appear also keeps changing, with a periodicity of 11 years. At the beginning of the cycle, sunspots appear at higher latitudes, approximately 30 to 40 degrees, both in the northern and the southern hemispheres. A sunspot typically lasts for about a month. After it disappears, succeeding sunspots appear at lower latitudes. In time, the sunspots appear close to the equator and eventually their density reduces to zero. This marks the end of the cycle. Sunspots then appear again at higher latitudes. This variation with latitude is shown in the upper plot of Figure 11.3, called the butterfly diagram.

A sunspot has a central dark region called umbra, surrounded by penumbra, where the intensity is higher than in the umbra. In the umbra, the magnetic field is vertical, perpendicular to the surface, whereas it is parallel to the surface in the penumbra. Sunspots occur in groups. A group has one large sunspot surrounded by several smaller ones. The size of the dominant sunspot is typically on the order of 30,000 Km.

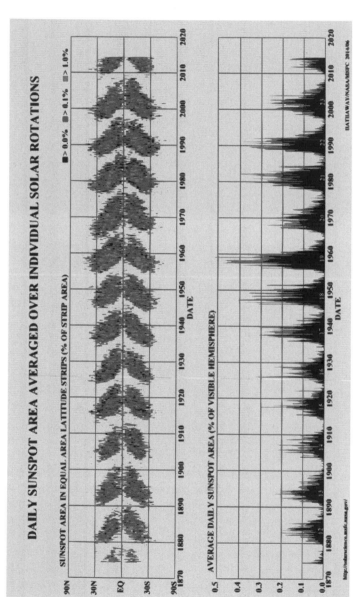

FIGURE 11.8: The upper figure shows the dependence of latitude of sunspots on time. The lower figure shows the percentage of total area covered by sunspots on the visible hemisphere of the Sun. At the beginning of the solar cycle, sunspots appear at higher latitudes, about 30° North and South. At this time, their number is small. After about 5.5 years, their numbers reach a maximum. At this time, they are found predominantly at about 15° North and South. Eventually after 11 years, their number becomes minimum. Now they are located close to the equator. (Image courtesy of Dr. David Hathaway, NASA, MSFC.)

The polarity of all dominant sunspots is the same in a particular hemisphere. The polarity is reversed in the opposite hemisphere. Furthermore, smaller sunspots have opposite polarity in comparison to the dominant spot. The polarities are reversed in the next 11-year cycle. For example, during a particular cycle, all the dominant spots in the northern hemisphere may be magnetic north, whereas those in the southern hemisphere may be magnetic south. In the next 11-year cycle, the dominant spots in the northern hemisphere would be magnetic south with opposite polarity in the southern hemisphere. Along with a change in the polarity of sunspots, the overall polarity of the Sun also undergoes a reversal. The large-scale magnetic field of the Sun shows a 22-year cycle.

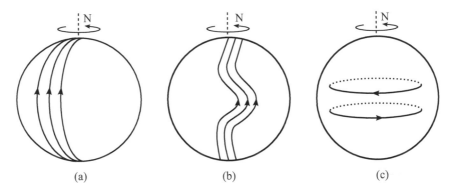

(a)   (b)   (c)

FIGURE 11.9: (a) The dipolar or poloidal magnetic field at the beginning of the solar cycle. (b) Due to higher angular velocity near the equator, the magnetic field starts to get distorted. (c) With time, the magnetic field lines become nearly parallel to the equator. Such a magnetic field is called toroidal.

At the beginning of the cycle, the large-scale magnetic field of the Sun is almost dipolar, with magnetic field lines going from geographic north to south poles, as shown in Figure 11.9. Such a field is called poloidal. The angular velocity of the Sun is higher near the equator in comparison to the poles, that is, the Sun does not rotate like a rigid body. However, plasma cannot move freely perpendicular to magnetic field lines because it consists of charged particles. Hence the magnetic field lines are frozen into the solar plasma and are dragged along with it. This leads to a distortion of the field lines, which now develop a component parallel to the latitudes. At this time the number of sunspots is near its minimum. A small number of sunspots start developing at high latitudes. As the magnetic field continues to distort, the number of sunspots reaches a maximum. At this point the large-scale magnetic is almost toroidal, that is, it circles around the Sun nearly parallel to the latitudes. Subsequently, the field again becomes nearly poloidal but with reversed polarity. The number of sunspots reaches a minimum. The mean latitude of sunspots at this stage is close to the equator. The large-scale field undergoes another reversal during the next 11 years. Hence after 22 years, the large-scale field returns to its original configuration, completing a full cycle. The sunspots form when the magnetic field in the solar interior gets pushed to the surface due to convective currents. Here we do not discuss the details of this process.

Besides the variation over the 11-year cycle, the number of sunspots also shows a long time variation. This is shown in Figure 11.10. We find that the number of sunspots is different for different cycles. The intensity of the Sun also shows a similar variation. The total solar flux is found to be large when the number of sunspots is large. One also sees some time intervals, such as the Maunder Minimum, where the number of sunspots is very small. During this time, the solar flux was also found to be significantly reduced. Historically, this coincides with a very cold period, the little ice age, on Earth. This also shows the significance of understanding the long-term variation of sunspots.

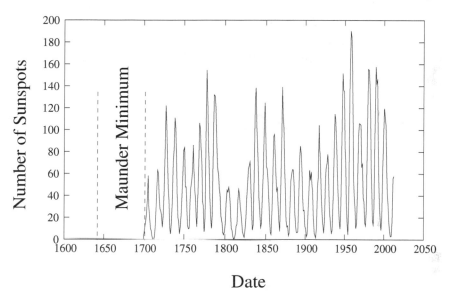

FIGURE 11.10: The time dependence of yearly averaged sunspots from 1700 to 2000. One sees long-term variation in the number of sunspots. The Maunder Minimum corresponds to the time interval 1645–1715. During this period, the number of sunspots observed was very small. (Data taken from SIDC-Solar Influences Data Analysis Center, Royal Observatory of Belgium.)

11.4   Some Transient Phenomena

Finally we introduce some transient phenomena associated with the magnetic field in the solar atmosphere. We observe several features that last only for a short time interval. One such feature is called *prominence* or *filament*. It is a region of ionized gas associated with the magnetic field in the solar atmosphere and typically lasts a few days or weeks. It is called a filament if viewed against the background of the solar disk. If viewed over a continuum range of visible frequencies, a filament appears darker than the surroundings. However, it appears bright when viewed at the Balmer H_α frequency. A similar feature is called a prominence, if viewed along the limb of the Sun, as shown in Figure 11.11. The figure shows a loop prominence. In this case, the prominence is produced by gas confined in a magnetic loop.

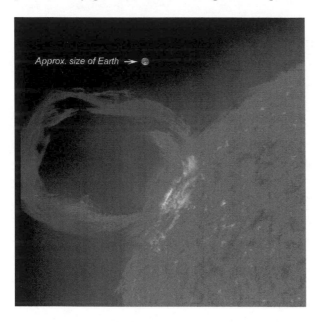

Approx. size of Earth

FIGURE 11.11: An image of prominence taken at extreme UV frequencies. (Figure courtesy of NASA.)

Another interesting solar phenomenon is solar flares. During a flare, a huge amount of energy stored in magnetic field in the solar atmosphere is released. The flares originate at the location of sunspots on the Sun. The time scale of a flare is typically between 1 second and 1 hour. During this short time interval, energy on the order of 10^{30} ergs may be released. This huge amount of emitted radiation causes considerable disturbances on Earth. In particular, it can affect communication, electronic equipment, as well as cause hazards to spacecraft. Hence it is important to make reliable predictions of their future occurrences.

Exercises

11.1 The diameter of a dominant sunspot is typically about 30,000 Km. Determine its angular diameter as seen by an observer on Earth. What percent of the solar disk is covered by this sunspot? Assume that it is located at the center of the disk.

11.2 The rotation period of the Sun near the equator is about 25 days and about 36 days near the poles. Determine the time taken by a point at the equator to complete one more revolution compared to a point near the pole.

11.3 Using Equation 8.69, determine the mean velocity of electrons and protons in the photosphere and the corona. Determine also the kinetic energies per unit volume in these two regions. Their temperature can taken to be 4,400K and 10^6K and the gas density, 10^{-7} g/cm^3 and 10^{-16} g/cm^3, respectively.

Chapter 12

The Solar System

The solar system is a gravitationally bound system consisting of the Sun, planets and their moons, smaller objects such as asteroids, minor planets, and comets, the interplanetary medium, and the hypothetical Oort cloud. It is a planar system with the orbits of all planets lying roughly in the same plane. The motion of each planet is dominated by the gravitational pull of the Sun. The effect of all other objects is relatively small and can be neglected to a good approximation. The Sun gives dominant contribution, about 99.86%, to the mass of the solar system. Its size, however, is much smaller in comparison to the radius of the solar system. The outermost planet, Neptune, is at a distance of about 30 AU, which is roughly 6,000 times the solar radius. Due to these large distances, the angular momentum of the solar system is dominated by the planets and not the Sun. The solar system actually extends to much larger distances. Beyond Neptune lies the Kuiper belt, at distance between 30 AU and 55 AU. It consists of a large number of small objects and its shape roughly resembles a donut. At much larger distances, on the order of 5,000 AU to 100,000 AU, lies the hypothetical Oort cloud. It is believed to be a spherically symmetric region, containing a huge number of small objects. The comets originate in the Kuiper belt and the Oort cloud.

The four inner planets, Mercury, Venus, Earth, and Mars, are called terrestrial planets due to their similarity to Earth. The remaining four, Jupiter, Saturn, Uranus, and Neptune, are called Jovian (Jupiter like) planets or gas giants. The mean radius, mass, sidereal period of rotation, and the inclination of the rotation axis of all the planets are summarized in Table 12.1. As can be seen from the table, there is a clear difference in the physical properties of these two classes. The Jovian planets have a much larger radius and mass. Furthermore, the terrestrial planets have a solid surface and composition similar

TABLE 12.1: The mean radius, mass, rotation period, and inclination of the rotation axis, relative to their revolution axis, for all eight planets. The negative sign in the sidereal periods of Venus and Uranus means that the direction of rotation of these planets is opposite to that of Earth.

Planet	Mean Radius (Km)	Mass (M_{Earth})	Sidereal Period of Rotation (days)	Inclination of Rotation Axis
Mercury	2439.7	0.055	58.6	$\sim 0°$
Venus	6051.8	0.866	−243	177.3°
Earth	6371.0	1	0.9973	23.44°
Mars	3396.2	0.107	1.026	25.19°
Jupiter	69,911	317.8	0.4135	3.13°
Saturn	58,232	95.159	0.444	26.73°
Uranus	25,362	14.536	−0.718	97.77°
Neptune	24,764	17.147	0.671	28.32°

to Earth. The Jovian planets, in contrast, are predominantly liquid. Excluding Mercury, all the planets have atmospheres.

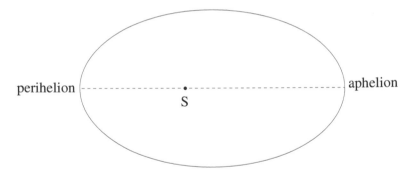

FIGURE 12.1: The elliptical orbit of a planet around the Sun (S). The closest point is called the perihelion, whereas the farthest is called aphelion. The eccentricities of all the planets are much smaller than that of the ellipse shown in this figure.

12.1   Orbital Properties of Planets

The orbits of all planets around the Sun are elliptical, with the Sun at one of the foci. As discussed in Chapter 5, their motion is described by Kepler's laws. The point at which a planet is closest to the Sun is called perihelion, the opposite point is called aphelion; see Figure 12.1. Some basic data on planet orbits is listed in Table 12.2. Mercury is the closest planet to the Sun and Neptune the farthest. The trajectories of planets in the solar system display some remarkable properties, as listed below:

- All planets revolve around the Sun in the same direction, from west to east.

- The rotation or spin of the Sun as well as most planets is also in the same direction as their direction of revolution, that is, from west to east. The only exceptions are Venus and Uranus.

- The orbits of all planets lie approximately in the same plane.

- The orbits are nearly circular. The eccentricity of the elliptical orbit is small.

This regularity suggests that the solar system may have formed by the collapse of a large rotating cloud of gas and dust. As we shall see, this hypothesis explains the main features of the solar system. The distances of planets from the Sun are given in Table 12.2. Here we give the semi-major axis of their orbit. These distances approximately follow an empirical rule called the Titius–Bode law. It states that the semi-major axis a of a planet in AU is given by

$$a = 0.4 + 0.3 \times 2^n ,$$

with $n = -\infty, 0, 1, 2...$ from Mercury to Neptune. The Titius–Bode law predicts the distances correctly except for a gap at $n = 3$. This may not be a failure as the largest object, Ceres, in the asteroid belt is found to lie at this position. The rule, however, does not predict the distance of Neptune reliably.

The time period of revolution of planets around the Sun is given by Kepler's law, Equation 5.9. Inserting the gravitational constant and the mass of the Sun, we obtain the time periods from Mercury to Neptune of, 0.24, 0.614, 1, 1.88, 11.87, 22.66, 84.33, and 165.19 years. These are in good agreement with observations.

The motion of all planets is governed primarily by the gravitational pull of the Sun, with the force exerted by remaining objects being relatively small. Furthermore, the orbit of natural satellites is governed by the planet around which it is in orbit. The effect of all other bodies is small and can be ignored to a good approximation. However, these bodies lead to small perturbations in the orbit. These are also measurable and must be included for an accurate description of the solar system. Furthermore, these perturbations, acting over a long period of time, can lead to large deviations. For example, the force exerted by the Moon (and the Sun) on the Earth leads to the precession of its rotation axis, which is a large effect for time intervals on the order of 10,000 years.

TABLE 12.2: The orbital parameters of planets, that is, the semi-major axis, eccentricity of the elliptical orbit of a planet, and the inclination of the orbital plane relative to the ecliptic.

Planet	Semi-major Axis a (AU)	Eccentricity	Orbital Inclination Relative to Ecliptic
Mercury	0.387	0.206	7.0°
Venus	0.723	0.007	3.4°
Earth	1.0	0.017	0°
Mars	1.524	0.093	1.85°
Jupiter	5.204	0.049	1.31°
Saturn	9.582	0.056	2.49°
Uranus	19.229	0.044	0.77°
Neptune	30.104	0.011	1.77°

The perturbations also lead to the remarkable phenomenon of resonance. Two different periodic motions are said to be in resonance when the ratio of their time periods, T_1 and T_2, is equal to a ratio of two small integers, that is,

$$\frac{T_1}{T_2} = \frac{n_1}{n_2}, \tag{12.1}$$

where n_1 and n_2 are small integers. A well-known example is the Moon, whose orbital period is nearly equal to its rotation period. This is an example of spin-orbit resonance. Due to this resonance, the Moon always presents the same face to an observer at Earth. A similar phenomenon is seen for the natural satellites of other planets. The sidereal rotation period of Mercury and its period of revolution are in the ratio 2/3. One also sees resonances in the time periods of revolution of planets. An example is the period of Pluto (246 years) and Neptune (165 years), which has the ratio 3/2. Another example is the moons (Ganymede, Europa, and Io) of Jupiter, whose periods are in the ratio 4:2:1. Furthermore, there are gaps in the asteroid belt called Kirkwood gaps. Some prominent examples of these gaps occur at distances of 2.5 , 2.82 and 3.27 AU from the Sun. The corresponding time periods are in the ratio, 1/3, 2/5, and 1/2, respectively, with respect to Jupiter's time period. These are explained in terms of the instability induced by Jupiter at these distances.

12.2   Retrograde Motion of Planets

The motion of a planet appears quite complicated if viewed from Earth. For example, at times they display retrograde loops, as shown in Figure 1.11, when the planet reverses its direction of motion, and makes a loop before returning to its original direction. Their uneven motion puzzled the ancient observers, who called them planets, which means wanderer in Greek. We can now explain this behavior in terms of the intrinsic motion of the Earth, as discussed in Chapter 1. In order to study this in more detail, let $\vec{r}_E$ and $\vec{r}_P$ represent the position vectors of Earth and another planet with respect to the Sun. The position vector of the planet with respect to Earth, $\vec{r}_{PE}$, is given by

$$\vec{r}_{PE} = \vec{r}_P - \vec{r}_E . \tag{12.2}$$

Assuming that the planets lie in the $x - y$ plane, the components of this position vector are given by

$$\begin{aligned} x &= r_P \cos \omega_P t - r_E \cos \omega_E t , \\ y &= r_P \sin \omega_P t - r_E \sin \omega_E t , \end{aligned} \tag{12.3}$$

where t is the time and ω_P and ω_E, respectively, represent the angular speeds of the planet and the Earth relative to the Sun. These formulae can be used to predict the position of any planet as observed from Earth. We will use them later in this section to determine the trajectory of Mars.

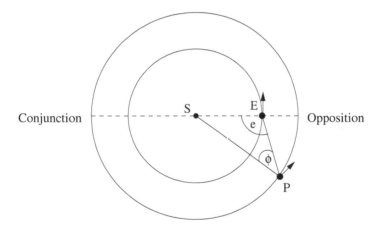

FIGURE 12.2: The orbit of Earth (E) and a superior planet (P) around the Sun (S). The positions corresponding to opposition and conjunction are indicated. The angles e and ϕ are, respectively, called the elongation and phase angle of the planet.

Planets that are further away from Sun, in comparison to Earth, are called superior; ones that are closer are called inferior. As shown in Figure 12.2, the elongation, e, of a planet is its angular position relative to the Sun, as observed from Earth. The phase angle, ϕ, is the angle of observation of the body relative to Sun. Let us first consider superior planets. They are said to be in *opposition* when they are on the opposite side of the Sun and in *conjunction* when on the same side, as illustrated in Figure 12.2. Due to the different angles of inclination of planes of revolution of different planets, a planet is never exactly in the opposite or same direction as the Sun. Hence an opposition is defined more precisely as the location when the RA of the planet differs from that of Sun by 180°. In any case, an opposition is the position when the planet is closest to Earth and conjunction when it is farthest. The retrograde loop happens when the planet is in opposition, as discussed in Chapter 1.

In Figure 12.3 we show the orbit of an inferior planet, such as Mercury or Venus. These planets are never in opposition. They are in inferior conjunction when they are closest to the Earth and superior conjunction when they are farthest, as shown in Figure 12.3. The maximum elongation, e, of Mercury and Venus is observed to be 28° and 47° respectively. Because they are always close to the Sun, they can only be seen for a certain period of time after sunset or before sunrise. In the middle of the night, they lie below the observer's horizon and hence are not observable. Furthermore, these planets show phases exactly as shown by the Moon. We observe the full planet when it is in superior conjunction. At this point, the planet is farthest from Earth and hence appears smallest. It shows a crescent phase when it is on the same side of the Sun as the Earth. Inferior planets also show retrograde motion. In this case, it also happens when they are closest to Earth, that is, at inferior conjunction.

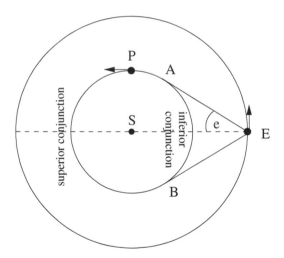

FIGURE 12.3: The orbit of Earth (E) and an inferior planet (P) around the Sun (S). The positions corresponding to superior and inferior conjunction are indicated. The positions A and B correspond to maximum eastern and western elongations, respectively. The angle e, shown in the figure, corresponds to maximum western elongation.

Let us now predict the time interval between two retrograde loops of a planet. Let us first consider a superior planet. Let T_{EP} be the time interval between two such events. This is also called the synodic period. In time T_{EP}, Earth completes one more complete revolution in comparison to the planet. Hence the angle traversed by Earth is 2π more than that traversed by the planet. This implies

$$\omega_E T_{EP} = \omega_P T_{EP} + 2\pi \,,$$

where $\omega_E = 2\pi/T_E$ and $\omega_P = 2\pi/T_P$ are the angular speeds of the Earth and the planet, respectively, and T_E and T_P the corresponding periods of revolution. This leads to

$$T_{EP} = \frac{T_E T_P}{T_P - T_E} \,. \tag{12.4}$$

Using this, the synodic period of Mars is found to be 2.14 years. For an inferior planet, the relationship is simply

$$T_{EP} = \frac{T_E T_P}{T_E - T_P} \,. \tag{12.5}$$

We next use Equation 12.3 to show the trajectory of Mars as observed by an observer on Earth. The observer moves along with Earth around the Sun but does not rotate along with Earth. Alternatively, we view the planet Mars in intervals of one sidereal day, after which the Earth rotates back to its original position relative to background stars. The resulting motion is shown in Figure 12.4. The retrograde loop is clearly seen when Mars is closest to Earth.

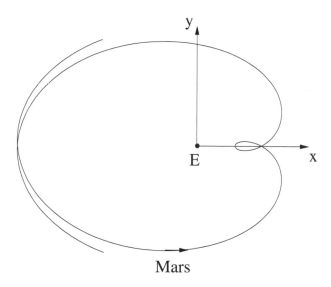

Mars

FIGURE 12.4: The orbit of Mars as seen from Earth (E). The retrograde loop is seen along the x-axis.

12.3   Albedo and Temperature of Planets

The planets are visible to us because they reflect part of the sunlight incident upon them. The remaining part is absorbed and maintains their temperature. They also emit blackbody radiation corresponding to their surface temperature. This radiation lies in the infrared part of the electromagnetic spectrum.

Let L_{Sun} denote the luminosity of the Sun. Consider a planet at distance r and radius R. The flux density at the surface of the planet is $L_{Sun}/(4\pi r^2)$. Hence the total flux incident on the surface of the planet, L_{in}, is given by

$$L_{in} = \frac{L_{Sun}}{4\pi r^2} \, \pi R^2 \,, \tag{12.6}$$

where πR^2 is the cross-sectional area of the planet. Let L_{ref} be the radiation reflected by the planet. The parameter Bond Albedo, A, is defined as the ratio of reflected to incident flux, that is,

$$A = \frac{L_{ref}}{L_{in}} \,. \tag{12.7}$$

This gives us an estimate of the total flux, integrated over frequencies and surface area, reflected by the surface. This is related to the observed brightness of the planet at visible frequencies. However, the radiation is not reflected isotropically. Hence the actual brightness perceived by an observer depends on the angle of observation. The Bond Albedo takes values between 0 and 1. For planets, Mercury has the smallest value of 0.068 and Venus the largest, 0.9. It is listed in Table 12.3 for all the planets.

It is also convenient to define the Geometric Albedo. This is defined as the ratio of the brightness of an object at zero phase angle to the brightness of a Lambertian surface, when both are illuminated by the same source. A Lambertian surface has the property that it reflects light isotropically. A rough, irregular surface is approximately Lambertian because the light after striking such a surface bounces off in all directions. The Geometric Albedo at visible frequencies is called the Visual Geometric Albedo. This is the quantity that is closely related to the observed brightness of planets when they are viewed at zero phase angle. For a superior planet, this would roughly be the case when the body is in opposition. For an inferior planet this would correspond to the position of superior conjunction. The Visual Geometric Albedo of planets ranges from 0.14 for Mercury to 0.67 for Venus and is listed in Table 12.3. Due to its high albedo and proximity to the Sun, Venus is the brightest object in the sky after the Sun and Moon. For Moon, the visual geometric albedo is 0.113.

In chapter 4, as an exercise, we computed the mean temperature of the Earth by assuming that it is a perfect blackbody. We now make a better estimate of the temperature of the Earth and other planets by taking into

TABLE 12.3: The albedo and temperature values for all the planets. The mean temperature refers to the observed mean surface temperature for terrestrial planets and the temperature at 1 bar pressure for Jovian planets.

Planet	Bond Albedo A	Visual Geometric Albedo	Blackbody Temp. (K)	Mean Temp. (K)
Mercury	0.068	0.14	440	440
Venus	0.9	0.67	184	737
Earth	0.31	0.37	254	288
Mars	0.25	0.17	210	210
Jupiter	0.34	0.52	110	165
Saturn	0.34	0.47	81	134
Uranus	0.30	0.51	57	76
Neptune	0.29	0.41	46	72

account their albedo. Let L_{Sun} be the total luminosity of the Sun. Consider a planet of radius R at a distance r. The total flux incident on the surface of the planet is given by Equation 12.6. Hence the rate at which energy is absorbed is equal to

$$L_{abs} = (1 - A) \frac{L_{Sun}}{4\pi r^2} \pi R^2 . \tag{12.8}$$

In steady state, this much energy is radiated by the planet. We assume that the emitted radiation corresponds to a blackbody temperature T. Equating L_{abs} with the luminosity of a blackbody $\sigma T^4 S$, where $S = 4\pi R^2$ is the total surface area, we obtain

$$T = \left[\frac{(1 - A)L_{Sun}}{16\pi\sigma r^2} \right]^{1/4} . \tag{12.9}$$

We call this the blackbody temperature of the planet. The computed values are shown in Table 12.3. For comparison, we also show the mean surface temperature for terrestrial planets. In general, the two cannot be compared directly due to the atmospheric greenhouse effect. The atmosphere does not allow the surface to radiate freely into outer space and hence raises the surface temperature. From Table 12.3 we see that for Mercury, which has no atmosphere, the agreement with blackbody temperature is very good. However, for Venus, the surface temperature is much higher. This is due to the greenhouse effect, which also accounts for the higher surface temperature of Earth. For the case of Jovian planets, Table 12.3 lists the observed temperature in their atmosphere at a pressure of 1 bar, which is equal to the atmospheric pressure at the Earth's surface.

12.4   Terrestrial Planets: Interior Structure

The study of planetary science has made remarkable progress since the 1960s by the use of space probes. The interior structure of Earth-like planets can be studied by surface probes that can observe seismic waves produced in the interior. The main features of Earth's interior are shown in Figure 12.5. The temperature and density steadily decreases from center to the surface. The mean density near the surface is about 3 gm/cm^3 and near the center, about 13 gm/cm^3. Starting from the center, we have the solid, metallic inner core, consisting of iron and other elements such as nickel and sulfur. Its temperature is very high, estimated to be about 6,000°C. This is followed by the outer core, which is liquid and has a composition similar to the inner core. The next layer is called the mantle, composed of very dense and hot silicate rock. Due to its high temperature, the rocky material can flow under high pressure. The outermost layer up to a depth of 35 Km from the surface is called the crust. It consists of light silicate rock. The upper layer of the mantle along with the crust is called the lithosphere. It consists of a large number of plates that float on the mantle. These tectonic plates undergo slow motion and occasionally collide with one another to produce mountains, earthquakes, etc. Furthermore, most volcanic activity occurs at the boundary of two plates. The hot magma that emerges as lava during volcanic eruptions originates in the upper mantle.

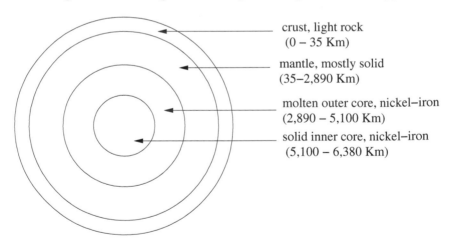

crust, light rock
(0 – 35 Km)

mantle, mostly solid
(35–2,890 Km)

molten outer core, nickel–iron
(2,890 – 5,100 Km)

solid inner core, nickel–iron
(5,100 – 6,380 Km)

FIGURE 12.5: The interior structure of Earth.

At the time of its formation, the Earth was very hot. With time, the surface has cooled off. The interior, however, is still very hot. Its high temperature cannot be explained in terms of the heat generated during its formation. There must be another heat source. It was speculated that a significant part of the heat generated in the Earth's interior is produced by radioactive decay of elements such as uranium, thorium, and potassium. This hypothesis was confirmed by laboratory detection of neutrinos produced in these decay reactions.

The Earth has an atmosphere that extends to several hundred kilometers above the Earth's surface. The outermost layer of the atmosphere is called the exosphere. It starts at an altitude of about 600 Km and eventually merges into outer space. In contrast to the lower layers, which are dominated by heavier gases, the exosphere is primarily composed of hydrogen and some helium. Another important component of the outer atmosphere is the ionosphere, which extends from about 100 Km to 600 Km. This region is ionized due to ultraviolet radiation from the Sun. The atmosphere is in hydrostatic equilibrium and hence described by Equation 8.7. Let $P(z)$, $T(z)$, and $\rho(z)$ be the pressure, temperature, and density at an altitude z above the Earth's surface. We have

$$\frac{dP}{dz} = -\rho(z)g,$$

where $g = GM/R^2$ is the acceleration due to gravity. It is reasonable to ignore the z dependence of g. Using the gas law and integrating, we obtain

$$P(z) = P_0 \exp\left[\int_0^z dz' \frac{\mu m_H g}{kT(z')}\right], \qquad (12.10)$$

where μ is the mean molecular weight of the atmosphere.

The Earth has a magnetic field, which, to a good approximation, has a dipole structure. The dipole axis of the field is inclined with respect to the Earth's rotation axis by about 11°. The magnetic field is maintained by the dynamo mechanism, as in the case of the Sun. In the present case, it is the rotational motion and convective currents of the liquid outer core which generate the magnetic field. The solar magnetic field flips its direction over a 22-year cycle. The magnetic field of Earth also undergoes reversals. However, in this case, the cycle has a much longer and irregular period, on the order of a million years.

The magnetic field strength at the surface of the Earth is about 0.5 Gauss. In the outer core it is estimated to be about 25 G. Because it is a dipolar field, it decays as $1/r^3$ with distance as we move out into space. The magnetic field plays a very important role for life on Earth. It protects Earth from the solar wind, which contains very energetic charged particles. In the absence of the magnetic field, these charged particles would strip away part of the Earth's atmosphere. This would lead to substantially larger flux of ultraviolet radiation on Earth's surface, which will not allow life to exist. The magnetic field deflects the solar wind away from Earth into outer space, as illustrated in Figure 11.6. This outer region of the Earth's magnetic field is called the magnetosphere. Some of the charged particles enter the Earth's atmosphere. They move parallel to the magnetic field lines and hence are deflected toward the poles. Due to collisions with oxygen and nitrogen atoms in the atmosphere, they produce bright colorful lights called the Aurora Borealis.

Other terrestrial planets also have an internal structure similar to Earth with a iron-nickel core, mantle, and crust. The relative size of these components, however, is different for different planets. Furthermore, the surface properties also vary significantly. The innermost planet, Mercury, has almost no atmosphere and its surface is heavily marked with craters. Hence it has an appearance closer to the Moon rather than Earth. Its rotation axis is nearly parallel to its axis of revolution. Hence, at all times, the flux density received from the Sun is maximum near the equator and decreases as we go toward the poles. The sidereal period of rotation (T_{sid}) is 58.65 days and the period of revolution (T) is 87.97 days. This is in the ratio $T/T_{sid} = 3/2$. This implies that the solar day of Mercury, $T_{solar} = 2T$. Hence one Mercury day is equal to 2 Mercury years. An observer at Mercury would experience daytime for a full year and nighttime during the next year. Due to the absence of atmosphere, the temperature, especially near the equator, is very high (~ 700K) during the day and low (~ 100K) during the night. Mercury has a weak magnetic field, with strength roughly 0.005 Gauss at surface. It has no natural satellite.

The absence of atmosphere can be explained by its proximity to the Sun and its small size. The mean temperature at the surface of Mercury is high due to its proximity to Sun. By the kinetic theory of gases, the temperature is related to the mean speed of the gas particles by Equation 8.69. Hence the mean speed is higher for higher temperatures. For sufficiently high temperature, the particles may have enough speed to escape the gravitational pull of Mercury, which is relatively small due to its small mass. Another important factor is the effect of the solar wind, which can partially strip the atmosphere and also raise its temperature. The solar wind strongly affects Mercury due to its proximity to the Sun and the weakness of its magnetic field. All these factors combined might explain the absence of Mercury's atmosphere.

Consider the atmosphere at a certain altitude, h, beyond which we assume that the density is so low that collisions among particles are almost negligible. At temperature T, a certain fraction of atoms or molecules would have speeds exceeding the escape velocity v_E. This fraction is given by

$$f_E = \int_{v_E}^{\infty} f(v)d^3v = 4\pi \int_{v_E}^{\infty} f(v)v^2 dv, \qquad (12.11)$$

where $f(v)$ is the Maxwell-Boltzmann distribution, Equation 8.67, and v is the total speed of a particle. Half of these particles would have their velocities directed outward from the planet and hence some would escape its gravitational attraction. If the rate at which the particles are escaping is sufficiently large, then by current time a large fraction would have left the planet.

These arguments also explain the absence of hydrogen and helium in the atmospheres of terrestrial planets. Recall that the Sun is predominantly composed of hydrogen and helium. The Jovian planets also have a very high proportion of these elements. Hence it is important to understand why they are absent in terrestrial planets. These elements are light and hence would be pushed up in the atmosphere, while the heavier gases settle down. You can show as an exercise that the hydrogen and helium would escape the Earth's atmosphere at a relatively large rate to account for their observed low density. The same argument applies to Venus and Mars.

We observe craters on the surface of many objects in the solar system that are caused by collision with meteorites. These are called impact craters. In general, all planets, moons, and asteroids are affected by this phenomenon. The largest craters seen in the solar system have diameters of a few thousand kilometers. In Figure 12.6 we show the cratered surface of Mercury near its south pole. Observations suggest a period of intense crater formation around 3.9 billion years. The craters formed since then have been preserved on the surface of Mercury, Mars, and some moons in the solar system. However, in the case of Earth, the surface gets modified due to geological activity, such as continental drift, earthquakes, volcanoes. Hence the craters are no longer visible.

FIGURE 12.6: Craters on the surface of Mercury near its south pole. (Image courtesy of NASA.)

Venus is the brightest astronomical object after the Sun and Moon, with a maximum apparent visual magnitude of −4.6. The planet Venus is closest to Earth in terms of size and mass. However, it differs considerably in surface and atmospheric properties. Its atmosphere is predominantly carbon dioxide. It is also very thick, with a surface pressure 92 times the atmospheric pressure at Earth. Due to the resulting greenhouse effect, the surface is very hot, with a mean temperature of 737K. The temperature does not change significantly between the equator and the poles, nor between day and night. The planet undergoes retrograde rotation at a very slow rate. Its rotation period is longer than its revolution period. Hence the length of a sidereal day is longer than a Venus year. Like Mercury, Venus does not have a moon. The surface of Venus is not visible from outside, being covered by thick clouds made of sulfur dioxide and droplets of sulfuric acid. These are highly opaque and reflective, and hence give rise to the high albedo and brightness of the planet. The surface of Venus contains a very high level of volcanic activity, much more than Earth. Overall, the conditions at the surface are very hostile. There is some evidence that suggests that in the past the planet may have been similar to Earth, containing oceans of water, but the conditions changed due to a runaway greenhouse effect. The magnetic field on Venus is very weak.

Mars is further away from the Sun in comparison to Earth and hence can be seen even in the middle of the night. Its maximum apparent visual magnitude is about −2.91, which is exceeded only by Jupiter and Venus. Its reddish, rust-like color is due to the presence of iron compounds on its surface, which presents a rocky, dusty, regolith structure. Its rotation period and inclination of the rotation angle are similar to Earth. Hence it displays seasons with the Sun being overhead at different positions at different times of the year. Its atmosphere is very thin and predominantly composed of carbon dioxide. The temperature varies steeply with time as well as from the equator to poles. The summer temperature at the equator varies from 180K to about 300K. At the poles the temperature may be as low as 130K. Hence it is a very cold planet. The magnetic field as well as plate tectonic activity are absent at present. However, geological evidence indicates these may have been present some billions of years ago. The Martian surface is also heavily cratered, especially in the southern hemisphere. As in the case of Mercury, this suggests that the surface has been preserved for billions of years since the time of heavy crater formation. Mars has two small moons, Phobos and Deimos.

12.5   Jovian Planets

The Jovian planets are much larger, more massive, and have a very different internal structure in comparison to the terrestrial planets. Their internal structure is depicted in Figure 12.7. The details vary between different planets. Jupiter and Saturn are composed predominantly of hydrogen and helium in liquid and gaseous form. The core of Jupiter is believed to be composed of rock and iron, whereas for other planets it is dominantly rock and ice. In the case of Jupiter and Saturn, the core is surrounded by a layer of liquid metallic hydrogen, which may also contain some helium. Metallic hydrogen means that some of the electrons in this layer are free and not bound to hydrogen atoms. This layer is surrounded by a layer of liquid molecular hydrogen followed by gaseous atmosphere, which is composed predominantly of molecular hydrogen and helium, similar in composition to the solar atmosphere. The internal structure of Uranus and Neptune is considerably different. It is estimated that the core is surrounded by a layer composed of water, ammonia, and methane ice. This is followed by an atmosphere consisting of hydrogen, helium, and methane gases. These planets do not have a solid surface. The visible surface lies in their atmosphere. In the case of Jupiter and Saturn, it is formed by clouds of ammonia crystals. For Uranus and Neptune it is due to the presence of methane gas in their atmosphere. Methane strongly absorbs red light and reflects or scatters blue light, thus leading to their observed bluish color. All the Jovian planets have a strong magnetic field.

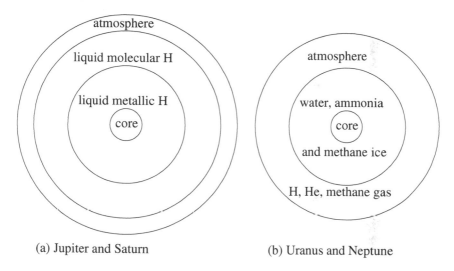

(a) Jupiter and Saturn   (b) Uranus and Neptune

FIGURE 12.7: The internal structure of Jovian planets.

Jupiter is often seen as the brightest object in the night sky, following the Moon. Venus is brighter but it is visible only for a limited time. The maximum apparent visual magnitude of Jupiter is -2.94. The atmosphere of Jupiter generates very strong storm systems with wind speeds greater than 400 Km/hour in some cases. Its rotation period, as well as that of all the Jovian planets, is shorter than that of Earth. The strong wind patterns may be due to its larger rotation period and radius. A prominent feature on its surface is the great red spot in its southern hemisphere. This is an anti-cyclonic storm, larger than the size of Earth, moving counterclockwise with a rotation period of about 6 days. It has been in existence for a very long time, possibly 400 years.

The atmospheric temperature of Jovian planets is very cold. The temperature at the position where the pressure is 1 bar, same as atmospheric pressure at Earth's surface, is given in Table. 12.3. The temperature increases as one moves toward the center. The temperature in the core of Jupiter may be as high as 24,000K. Due to their low temperature and strong gravitational pull, the Jovian planets are able to retain hydrogen and helium in their atmosphere.

The most striking observational feature of Saturn is its rings, which can be seen even with a small telescope. They lie in its equatorial plane with the innermost ring starting at a distance of 66,900 Km from the center of Saturn. They extend to very large distances of more than 200,000 Km. Their thickness is very small, less than 1 Km and in some regions as small as 10 m. They are composed of dusty water ice particles of varying sizes, the largest being the size of a large boulder, about 10 m. The innermost ring is called the D ring. This is followed by C, B, and A, which are the main rings of the planet. The rings orbit around Saturn at different angular speeds. We observe several gaps, the largest being 4,800 Km wide. It is called the Cassini division and separates the B and A rings. The rings are probably formed out of primordial matter that could not condense into a satellite due to tidal forces exerted by Saturn and its moons. Alternatively, a moon might have disintegrated as it entered the Roche limit of the planet. The gaps probably arise due to the destabilizing effect caused by the moons of Saturn, similar to the resonant phenomenon that causes gaps in the asteroid belt. All the other Jovian planets also have rings but these are not as prominent as in the case of Saturn.

All the Jovian planets have a large number of natural satellites. Jupiter has 67 known moons, Saturn 53, Uranus 27, and Neptune 13. The largest moons of Jupiter, in order of their increasing distance from the planet, are Io, Europa, Ganymede, and Callisto. Ganymede is the largest moon in the entire solar system. As we have already mentioned, Ganymede, Europa, and Io show resonant period of revolution around Jupiter. Furthermore, like our moon, their orbital period is also in resonance with their rotation period. Hence they always maintain the same face toward Jupiter. Saturn has 16 moons that display this phenomenon. The largest moon of Saturn is Titan. It is the only moon in the solar system known to have clouds and a dense atmosphere, similar to a planet.

12.6   The Moon

Our Moon displays some rather special properties. It is the largest satellite among the terrestrial planets. Mars is the only other terrestrial planet that has satellites. However, both of its satellites, Deimos and Phobos, are tiny in comparison with the Moon. Their mean radii are 6 and 11 Km, respectively. The Jovian planets, in contrast, have a large number of satellites, as discussed in the previous section. It is very likely that the process of formation of our Moon is different from that of other natural satellites. In most cases, the satellites are captured by the planets from the planetary disk. The Earth's Moon, however, might have been formed due to a collision of a very massive object with Earth.

The Moon has a highly cratered surface, very thin atmosphere, and very weak magnetic field. It has a nearly circular orbit around the Earth with an eccentricity of about 0.05. Its orbit is also significantly affected by the Sun's gravitational field. Hence we cannot determine its trajectory precisely by simply considering the force due to Earth. Its direction of revolution around the Earth is the same as that of Earth's around the Sun. Furthermore, the orbit is nearly parallel to the ecliptic plane, with an inclination of about 5°. The orbital plane also undergoes slow change due to the perturbations caused by the Sun and Earth. Its period of rotation is almost the same as its period of revolution. Hence one of its faces always remains hidden to an observer on Earth.

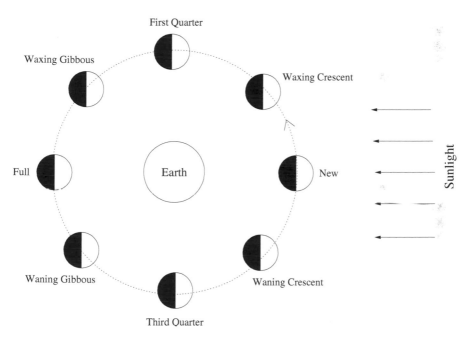

FIGURE 12.8: The different phases of the Moon, as observed from Earth. The Moon moves from west to east, with its orbit inclined at approximately 5° to the ecliptic plane.

The different phases of the Moon are illustrated in Figure 12.8. These phases arise due to its relative position with respect to the Sun. The Moon rises from the east and sets in the west due to Earth rotation. Its revolution around the Earth is from west to east. Relative to fixed stars, it returns to its original position after approximately 27.3 days. This is called the sidereal period. Its synodic period, that is, the period over which it completes one full cycle of phases, is about 29.5 days. The difference in these periods arises because the Earth travels approximately 1/12 of its orbit in one month. Hence the Moon has to travel an additional 1/12 of its orbit in order to return to its original position relative to the Sun and the Earth.

At its new moon phase, the Moon is in the same direction, that is, has the same longitude, as the Sun. However, its latitude differs slightly due to the tilt of its orbit relative to the ecliptic plane. The fraction of the Moon that appears bright starts to increase beyond the new moon. At waxing crescent, its elevation is maximum at a particular position on Earth at 3 PM. At other positions on Earth, it may be visible in the evening, at a lower elevation, or at night, close to the horizon. At first quarter, we observe the half moon, followed by waxing gibbous, and full moon as shown in Figure 12.8. At full moon, its elevation is maximum at midnight. Beyond this, the fraction of its illuminated region begins to decrease.

The position of the Moon as viewed from Earth depends on its orbital motion as well as the rotation of Earth. Because the direction of revolution of Moon is same as the rotation of Earth, the position of the Moon shifts eastward by about 12° every day. In time units, 12° is about 50 minutes. Hence the moon rise is delayed by 50 minutes every day.

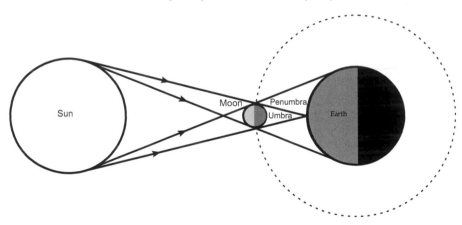

FIGURE 12.9: A solar eclipse or occultation occurs when the shadow of the Moon falls on the Earth.

12.6.1   Eclipses and Occultations

The fact that the angular size of the Moon is almost identical to that of the Sun leads to the fascinating phenomena of eclipses, as shown in Figures 12.9 and 12.10. A solar eclipse or occultation occurs when the Moon lies between the Earth and the Sun. This configuration corresponds to the new moon phase.

The orbital plane of the Moon is inclined at an angle of about 5° to the ecliptic plane. Had the two planes been exactly parallel, an eclipse would have occurred every month. However, due to the inclination, an eclipse can occur only when the Earth, Moon, and Sun lie along the line of intersection of the two planes. Hence the phenomenon is not very frequent. The central dark region of the shadow on Earth is called umbra. The surrounding region, which is partially in shade, is called penumbra. The lunar eclipse occurs when the Earth lies between the Sun and the Moon. This is possible only when the Moon is in its full moon phase. There are many other types of eclipses or occultations that play an important role in astronomy. For example, the Moon may occult a star or another planet. Another important example is the occultation of a star by a planet.

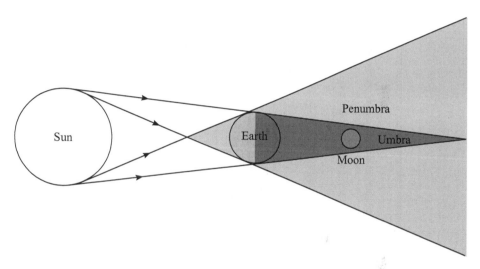

FIGURE 12.10: The lunar eclipse occurs when the shadow of the Earth falls on the Moon.

12.7 Why Did Pluto Lose Its Planetship?

Pluto was originally classified as a planet. It is composed primarily of rock and ice. It lies in the Kuiper belt and has a highly eccentric orbit, with its distance from Sun ranging from 30 to 49 AU. It is relatively small, with a mass 500 times smaller than that of Earth. However, by the year 2006, astronomers had found other objects of size similar to that of Pluto. It was impractical to classify them all as planets. Instead, in 2006 the International Astronomical

Union (IAU) decided to make a more precise definition of a planet. A planet is now defined as a celestial body that

1. Is in orbit around the Sun

2. Has sufficient mass for its self-gravity to overcome rigid body forces so that it assumes a hydrostatic equilibrium (nearly round) shape

3. Has cleared the neighborhood around its orbit

Pluto does not meet criteria 3. Hence it was reclassified as a dwarf planet.

12.8   Formation of the Solar System

The solar system is formed due to the contraction of a primordial cloud of gas and dust called a nebula, as discussed in Chapter 10. As the cloud shrinks, the central region becomes dense and opaque to form a protostar, which eventually acquires hydrostatic equilibrium. The present best estimate of the age of the solar system is about 4.6 billion years. Let us assume that the nebula initially had a small angular speed. As it collapses, its rotational speed increases due to conservation of angular momentum, leading to a large angular speed of the protostar. The surrounding material continues to fall onto the central object. The disk forms due to the centrifugal force that is present if the system is analyzed in a rotating frame. For simplicity, let us assume that the rotational speed of the protostar is independent of the angular position. In this case, the centrifugal force is maximum near the equator and negligible near the poles. Hence the material near the center would be pulled away from the center, forming a disk, as shown in Figure 12.11. All the planets form in this disk and hence lie approximately in the same plane. Furthermore, they acquire their angular momentum from the disk and hence all of them revolve in the same direction.

The matter in the protostar and the surrounding disk also dissipates energy due to friction. This dissipative process, however, conserves angular momentum. This leads to a further enhancement of the disk. This is because, as a particle loses energy, it has to move to larger distances in order to maintain its angular momentum. This process essentially leads to a transfer of angular momentum to larger distances from the center. Hence it provides an explanation for why most of the angular momentum of the solar system resides in the planetary disk.

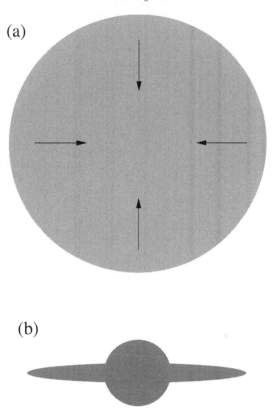

FIGURE 12.11: A primordial cloud of dust and gas collapses to form the Sun and the protoplanetary disk. The planets and asteroids eventually form in this disk.

The disk was initially very hot and in the gaseous state. As it cools, it starts to condense much like the condensation of raindrops from water vapor in the atmosphere. Heavier particles condense at higher temperatures. Hence silica and iron compounds are the first to undergo condensation. These droplets grow in size as more material condenses on their surface. They may also collide and undergo fusion with other droplets. This process leads to a large number of asteroid size objects, called planetesimals, which must have populated the disk at an early stage. Some of these objects might become sufficiently large so that they may attract other objects by their gravitational pull. By capturing all the matter in their vicinity, the planetesimals eventually form planet size objects.

The formation of Jovian planets starts earlier because the temperature is lower at larger distances from the protostar. As the temperature cools further, lighter compounds (hydrogen, ammonia, water, methane) condense, forming the outer layers of the Jovian planets. Eventually, as the Sun enters the main sequence, the solar wind clears out the interplanetary medium, thus halting further development of planets. In the case of terrestrial planets, condensation of lighter compounds such as ammonia never takes place. Hence they never become as large as the Jovian planets.

Exercises

12.1 Determine the orbital angular momentum of the four Jovian planets. Compare this with the angular momentum of the Sun. Computing the Sun's angular momentum is a little complicated because it does not rotate as a rigid sphere. However, for simplicity, we can assume that it is a rigid sphere of uniform density and its mean rotational time period is 30 days.

12.2 Use Equation 5.9 to determine the time period of rotation of the Moon around the Earth. Its semi-major axis is 384,399 Km. (Ans: 27.46 days)

12.3 Verify that the time periods of objects located at the Kirkwood gaps at distances, 2.5, 2.82, and 3.27 AU, from the Sun are in the ratio 1/3, 2/5, and 1/2, respectively, with respect to the time period of Jupiter.

12.4 Verify Equation 12.5 for the synodic period of an inferior planet. Use Equations 12.4 and 12.5 to determine the synodic period of all the planets.

12.5 Determine the time interval of the retrograde loop of Mars. To simplify the calculation, assume the configuration shown in Figure 12.4. Here Mars is in opposition, closest to Earth, at $t = 0$. The retrograde loop begins at some $t < 0$ when $y = 0$ and ends at $t > 0$ when y is again 0. Hence we need to set

$$y = r_P \sin \omega_P t - r_E \sin \omega_E t = 0 .$$

In order to get an analytic expression, expand the sin functions, using

$$\sin(\theta) = \theta - \frac{\theta^3}{6} + ... ,$$

ignoring higher order terms. This is valid for small θ. Show that the time t for which $y = 0$ corresponds to

$$t^2 = \frac{6(r_P \omega_P - r_E \omega_E)}{r_P \omega_P^3 - r_E \omega_E^3} .$$

Using this, show that the retrograde loop starts at time $t \approx -0.193$ years and ends at $t \approx 0.193$ years.

12.6 Determine the solar day of Mercury using the data for sidereal day and period of revolution. Repeat this exercise for Venus. Note that Venus undergoes retrograde rotation. (Ans: 1 solar day = 2 Mercury years).

12.7 Use Equation 12.11 to approximately estimate the fraction of particles that have speeds greater than the escape speed v_e at temperature T. Using the Maxwell–Boltzmann distribution, we obtain

$$f_E = \left(\frac{m}{2\pi kT}\right)^{3/2} 4\pi \int_{v_E}^{\infty} dv v^2 e^{-mv^2/2kT}.$$

We make a rough order of magnitude estimate of this integral by replacing the v^2 factor inside the integrand by vv_E. This gives

$$f_E \sim \sqrt{\frac{2m}{\pi kT}} v_E e^{-mv_E^2/2kT}.$$

Estimate the fraction of H_2 molecules that have $v > v_E$ at a temperature $T \sim 1,000K$, corresponding to temperature at the base of the exosphere, located at an altitude of 500 Km. (Ans: approx. 10^{-6})

12.8 Use the result of Exercise 12.7 to roughly estimate the rate at which H_2 can escape the Earth's atmosphere. Consider the hydrogen molecules at an altitude of 500 Km. These particles would escape in a time interval required to travel a distance over which the density of atmosphere changes appreciably. Assume this distance to be 500 Km. Find the corresponding time, assuming that they move at speed v_E. Hence find the order of magnitude of the time interval over which all the hydrogen gas would escape the Earth's gravitational pull. The reader can also verify that the corresponding time scale for oxygen or nitrogen molecules is extremely large. (Ans: 45 sec, about 1 year)

Chapter 13

Binary Stars

Most stars are observed to lie in clusters, that is, a collection of stars gravitationally bound to one another. We observe a wide array of clusters, from binary systems to globular clusters, composed of hundreds of thousands of stars. In a binary system, the two stars move in elliptical orbits with respect to their common center of mass. Viewed from each star, the other star would also appear to move in an elliptical orbit. The binary star systems are particularly important because they allow determination of the masses of the binary partners. One observes a wide range of binary systems with periods ranging from a few days to hundreds or thousands of years. In some cases the two stars nearly touch one another, whereas in other systems they might be separated by a distance greater than 100 AU.

Binary star systems are identified by the periodic variation of their position, velocity, spectra, or brightness. In some special circumstances, it is possible to observe both binary partners. In most cases this is not possible and the binary nature of the system must be deduced indirectly. We point out that two stars that might appear very close in their angular separations need not form a binary system. They might be widely separated from one another along the radial direction and hence there may be no association between them. The binary nature is identified by a periodic change in their properties.

13.1 Kinematics of a Binary Star System

A system of two gravitationally bound point objects can be solved analytically, as described in Chapter 5. Here we apply this formalism to a binary star system assuming that the two stars are very close to one another so that the gravitational pull of other stars does not a have significant effect on their motion relative to one another. An observer on either of these stars sees the other star moving in an elliptical orbit, described by

$$r = \frac{r_0}{1 - \epsilon \cos\theta}, \tag{13.1}$$

where r and θ are the plane polar coordinates in the plane of the ellipse. The eccentricity of the ellipse, ϵ, and r_0 are constants. An example of a binary system is illustrated in Figure 13.1. Here we show the motion of two stars A and B whose masses M_A and M_B, are in the ratio $M_A/M_B = 2$. The time period T of the orbit is given by (Equation 5.9),

$$T^2 = \frac{4\pi^2 a^3}{(M_A + M_B)G}, \tag{13.2}$$

where $2a$ is the length of the major axis of the orbit of B about A.

With respect to their common center of mass, C in Figure 13.1, the two stars also go in an elliptical orbit with the same ϵ but different r_0. Hence the major axis of their ellipses is different. Depending on the eccentricity, the orbit of one of the stars may be contained entirely inside the orbit of the partner, as in the case of Figure 13.1.

We next determine the motion of individual stars with respect to their common center of mass. Let us assume that the center of mass of the two stars lies at the origin of our coordinate system. Let $\vec{r}_A$ and $\vec{r}_B$ denote the positions of the two stars. We denote the position of B relative to A by $\vec{r}$, that is,

$$\vec{r} = \vec{r}_B - \vec{r}_A. \tag{13.3}$$

In Figure 13.1, the magnitudes of the vectors $\vec{r}$, $\vec{r}_A$, and $\vec{r}_B$ are denoted by symbols L, L_A, and L_B at time t_1. We can write this vector as

$$\vec{r} = r\cos\theta\hat{x} + r\sin\theta\hat{y}, \tag{13.4}$$

in terms of the plane polar coordinates (r, θ). Furthermore,

$$M_B\vec{r}_B + M_A\vec{r}_A = 0, \tag{13.5}$$

as the center of mass is located at the origin. The two equations, Equations 13.3 and 13.5, imply that

$$\begin{aligned} \vec{r}_A &= -\vec{r}\,\frac{M_B}{M_A + M_B}, \\ \vec{r}_B &= \vec{r}\,\frac{M_A}{M_A + M_B}. \end{aligned} \tag{13.6}$$

These equations show that the trajectory of the two stars about the center of mass C of the binary is also elliptical, with the distance $r_B/r_A = M_A/M_B$. We can express these as

$$\begin{aligned} r_A &= \frac{r_{0A}}{1 - \epsilon\cos\theta}, \\ r_B &= \frac{r_{0B}}{1 - \epsilon\cos\theta}, \end{aligned} \tag{13.7}$$

where the eccentricities are the same and the ratio

$$\frac{r_{0A}}{r_{0B}} = \frac{M_B}{M_A}. \tag{13.8}$$

Let $2a_A$ and $2a_B$, respectively, denote the lengths of the major axis of the elliptical orbits of A and B about C. Their ratio is given by

$$\frac{a_A}{a_B} = \frac{M_B}{M_A}. \tag{13.9}$$

Furthermore, $r_0 = r_{A0} + r_{B0}$, which implies that

$$a_A + a_B = a. \tag{13.10}$$

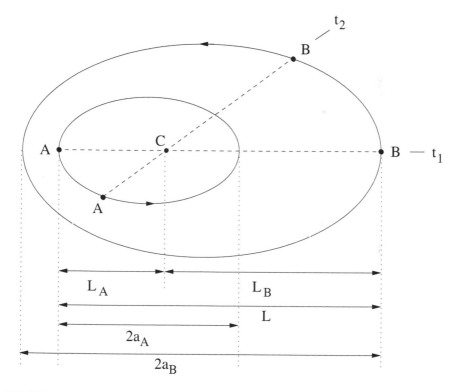

FIGURE 13.1: The motion of two stars, A and B, under their mutual gravitational attraction. Each star orbits around their center of mass C in an elliptical orbit. Here their masses, M_A and M_B respectively, are taken such that $M_A = 2M_B$. The positions of the two stars at two different times t_1 and t_2 are shown. At time t_1, the two stars lie at the two ends of the dashed horizontal line. As viewed from C, the stars A and B are located at distances L_A and L_B, respectively, at this time. The lengths of the major axes of the orbits of A and B around C are $2a_A$ and $2a_B$, respectively.

In general, the orbital plane of a binary may not be perpendicular to the line of sight. Let η be the inclination angle between the line of sight and the unit vector, $\hat{n}$, perpendicular to the plane of their trajectory. The observer will see the trajectories of the two stars, projected onto a plane perpendicular to the line of sight. The projected trajectory is also an ellipse, whose properties, such as major and minor axes, eccentricity, are different from that of the real trajectory.

The center of mass of a binary system moves in space due to the proper motion of stars. Let us assume that it moves in a fixed direction with constant speed. The resulting motion of the two stars is shown in Figure 13.2 for a particular choice of parameters. Hence both stars show periodic change in position superimposed on a linear motion in some direction.

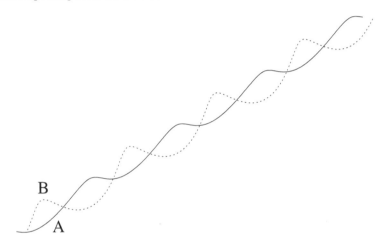

FIGURE 13.2: The motion of a binary star system, A and B, across the sky. The parameters are taken to be same as in Figure 13.1. The angle of inclination is taken to be zero. The center of mass of the system moves in some fixed direction due to proper motion. The oscillations arise due to the motion of each star with respect to the center of mass.

13.2 Classification of Binary Stars

Binary stars are divided into various categories, depending on their observational properties. The brighter star in a binary system is called the **primary** and the dimmer, the **secondary**. If we can observe both stars, the resulting system is called a **visual binary**. This is possible only in a small fraction of such systems. The angular separation between the two stars must be sufficiently large so that they can be resolved. In some cases the secondary is too dim to be observed. If we can observe only one of the stars in a binary system, then the resulting system is called an **astrometric binary**. In this case the binary nature is deduced through the period changes in the position of the observed star. Let us assume that in the example shown in Figure 13.2, star B is not directly observable. The periodic motion of A, however, still allows us to determine the period of the binary system.

In many cases we may not be able to directly observe the shift in position of either of the stars. The binary nature is revealed by the periodic shift in the spectral lines or the observed flux. The shift in spectral lines occurs due to the Doppler effect as the two stars move back and forth along the line of sight. Such a system is called a **spectroscopic binary**. If the spectrum of both stars is observable, it is called a double-lined spectroscopic binary, otherwise it is a single-lined spectroscopic binary.

A binary system that shows periodic variation of its total flux is called a **photometric binary**. The variation is many cases arises when the two stars periodically eclipse one another. These are called **eclipsing binaries**. Let us first assume that the distance between stars is much larger than their radii. In such cases, eclipses can happen only if the plane of the orbit is nearly parallel to the line of sight, corresponding to an inclination angle of approximately 90°. Hence, in this case, we are able to deduce the inclination angle of the system. The luminosity of the system remains roughly constant for most of the period but shows a sudden dip when the one of the stars eclipses the other. The shape of the light curve depends on whether the eclipse is total or partial. In the case of a total eclipse, one of the stars is completely eclipsed by the other, larger star for a certain period of time. Hence, at a minimum, the light curve may remain constant for a certain period of time, as shown in Figure 13.3. One also sees a secondary eclipse when the larger star is partially eclipsed by the smaller star. If the orbit is circular, the time interval τ between a primary and the adjacent secondary eclipse is equal to $T/2$. The situation changes considerably if the two stars are very close to one another. In such cases the shape of a star may be distorted into an ellipsoidal due to the presence of the binary partner. As this star revolves around the partner, we see its different faces, which vary in brightness. In this case the brightness would show a time dependence even if there is no eclipse.

Depending on the nature of the light curve, the photometric binaries are further classified as Algol stars, β Lyrae stars, and W Ursae Majoris (W UMa) stars. The flux variation for Algol type stars is caused primarily by eclipses. The typical light curve is shown in Figure 13.3. The brightness remains approximately constant between the eclipses. Furthermore, it is possible to clearly identify the time interval over which the eclipse lasts. In some cases the secondary eclipse may not be visible. In β Lyrae stars and W UMa stars, the distortion in shape is also responsible for the variation in brightness. Hence the brightness varies even outside the eclipses and it not possible to identify the beginning and end of the eclipses. For β Lyrae we observe the secondary minima in all cases. In the case of W UMa stars, the difference in depth between the different minima is negligible.

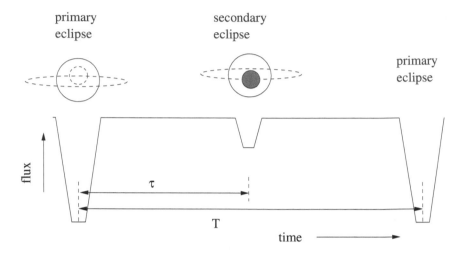

FIGURE 13.3: The light curve of an eclipsing binary star. The flux falls dramatically during the primary eclipse, when the primary star is eclipsed by the secondary. One also sees a smaller dip when the secondary is eclipsed by a primary.

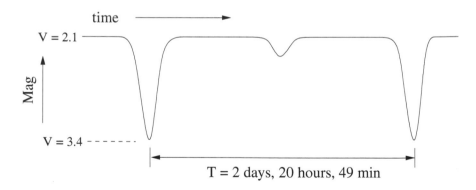

FIGURE 13.4: The light curve of the Algol system. The apparent visual magnitude, V, falls from 2.1 to 3.4 during the primary eclipse. The time period of the system is 2 days, 20 hours, and 49 minutes.

Some prominent examples of binary systems are

- Castor: This was the first binary star system discovered. The discovery was made by William Herschel around 1800. It is located at a distance of about 15.6 pc in constellation Gemini; Figure 1.8. The orbital period of this system is about 467 years, and the angular separation between the two stars about 6″. Hence, it has still not completed one orbit since its first observation as a binary system. Later, each star in this system was found to be a spectroscopic binary. Furthermore, it is bound to another faint binary system at an angular separation of about 72″, making it a sextuplet system.

- Sirius A and B: Sirius A is the brightest star in the sky. Its binary partner, Sirius B, is a white dwarf, which can be seen only with a powerful telescope. The existence of Sirius B was first deduced in 1844 by the periodic variations in the position of Sirius A. Hence at the time of its discovery, this system was an astrometric binary. The binary partner, Sirius B, was discovered a few decades later, making this system a visual binary. The time period of the system is 50.1 years.

- Mizar and Alcor: These two stars are very close to one another in the handle of Ursa Major (see Figure 1.9). Mizar, the brighter of the two stars, is itself a visual binary composed of Mizar A and Mizar B. Both Mizar A and B are themselves spectroscopic binaries. In 1997, astronomers were able to resolve the two stars in the Mizar A binary system by interferometric measurements. Alcor is also an astrometric binary. It has been suggested that the Alcor and Mizar systems might be gravitationally bound to one another. If confirmed, this would make this also a sextuplet system.

- Algol A and B: This is an eclipsing binary in the constellation Perseus with a time period of about 2.9 days. The visual magnitude drops from 2.1 to 3.4 during the primary eclipse. The light curve is schematically shown in Figure 13.4. This was the first eclipsing binary observed and has fascinated astronomers for hundreds of years due to its rapid change in flux. This rapid change also leads to its name, which is derived from an Arabic word which means "The Demon's Head." The two stars, Algol A and B, are very close to one another and cannot be resolved. From the light curve it has been deduced that Algol A is a very hot and bright star of spectral type B, whereas Algol B is a little larger and much cooler and dimmer K type star. The Algol variables are named after this system. This system also presents a paradox because Algol A is more massive than B but still on the main sequence, whereas Algol B has already entered the giant phase. This paradox has been resolved by arguing that Algol A lost a part of its mass to its partner at a certain stage in its evolution. We discuss this later in this chapter.

13.3 Mass Determination

A major advantage of a binary system is that, in certain circumstances, it is possible to determine the masses of the two stars. For example, consider a visual binary at zero angle of inclination. By separating out its proper motion from the oscillatory motion, we can determine the elliptical orbit of each star relative to the center of mass. This allows measurement of the major axis parameters, $2a_A$, $2a_B$, and $2a$, of the two stars A and B (Figure 13.1). The time period, T, can be measured for all binaries. Hence, using Equation 13.2, we can determine $(M_A + M_B)$, the sum of the masses of the two stars. Furthermore, using Equation 13.9, we can obtain the ratio of the masses M_B/M_A. Knowing their sum and ratio allows us to determine both M_A and M_B.

One complication that arises in the above analysis is that the binaries are usually observed at a non-zero angle of inclination. For visual binaries it is possible to determine this angle and hence deduce the true trajectory of the stars. This deduction relies on the fact that, in a binary system, the trajectory of star B, with respect to star A, is an ellipse with star A at one of its foci. Here we explain the basic idea with a simple example.

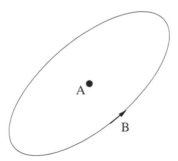

FIGURE 13.5: The observed orbit of star B with respect to star A. In this case, star A is located at the center of the ellipse, suggesting that the true orbit is circular.

Let us assume that Figure 13.5 represents the observed orbit of B relative to A. Here star A is at the center of the ellipse traversed by B. Because A must be at one of the foci, this implies that the true orbit must be a circle with diameter equal to the major axis, $2a$, of the ellipse. Let $2b$ represent the minor axes. The tilt angle η, discussed in more detail in Exercise 13.4, can be determined by the relationship $b = a \cos \eta$. With this information, the orbits of both particles about the binary center of mass can be deduced and used to determine their masses. In general, the true orbit is an ellipse and the observed trajectory may appear more complicated. The star A may not even lie along the major axis of the observed ellipse. Determining the tilt in these cases is more complicated and we do not go into such details here. In the case of astrometric binaries, it is not possible to determine the tilt angle. Hence we can only determine the sum of the masses, not the individual masses of the two stars.

We next discuss spectroscopic binaries. Let $\vec{v}_A$ and $\vec{v}_B$ denote the orbital velocities of two stars. By Doppler shift measurements, we can determine their components along the line of sight, as well as the time period T. It is not possible to determine the angle of inclination of the orbit. Using this information we can extract some information about the mass of the binary. The general case of elliptical orbit is complicated. Here we simplify our analysis by assuming that the true orbit is circular. In this case the speeds of the two stars, v_A and v_B, do not change with time. Let v_{Ar} denote the maximum value of the component of $\vec{v}_A$ along the line of sight. This component is positive when star A is moving toward the observer. At a certain instant, this component takes its maximum value, as discussed in Exercise 13.4. At the same instant, star B has maximum speed directed outward along the line of sight. We denote its magnitude, that is, absolute value, by v_{Br}. These components are given by (see Exercise 13.4),

$$
\begin{aligned}
v_{Ar} &= v_A \sin \eta, \\
v_{Br} &= v_B \sin \eta.
\end{aligned}
\tag{13.11}
$$

Differentiating Equation 13.5, we find, at any instant,

$$M_B \vec{v}_B + M_A \vec{v}_A = 0. \tag{13.12}$$

This implies that

$$\frac{M_B}{M_A} = \frac{v_{Ar}}{v_{Br}}. \tag{13.13}$$

The time period of the binary is related to orbital speed v_A by

$$v_A = \frac{2\pi a_A}{T}. \tag{13.14}$$

This leads to

$$v_{Ar} = \frac{2\pi a_A}{T} \sin \eta, \tag{13.15}$$

along with a similar equation for v_{Br}. Using these as well as Equations 13.10 and 13.13, we obtain

$$a = \frac{v_{Ar} + v_{Br}}{\sin \eta} \frac{T}{2\pi}. \tag{13.16}$$

Substituting this in Equation 13.2, we obtain

$$(M_A + M_B) \sin^3 \eta = (v_{Ar} + v_{Br})^3 \frac{T}{2\pi G}. \tag{13.17}$$

This, along with Equation 13.13, allows determination of $M_A \sin^3 \eta$, $M_B \sin^3 \eta$.

The masses cannot be determined because the inclination angle is unknown. For single-lined binaries, one can only determine the speed of one of the stars. Hence only a certain combination of the masses and the inclination angle can be determined.

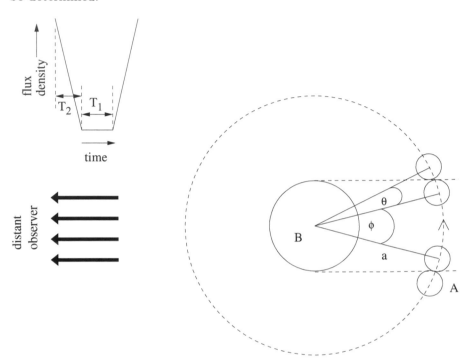

FIGURE 13.6: The position of star A at various stages during the primary eclipse of the binary system. The center of star A traverses an angle ϕ about the center of star B for the duration of the total eclipse. It traverses an angle θ as it moves out of the occultation. The corresponding flux density seen by a distant observer, from the beginning to the end of the eclipse, is also shown. The time periods T_1 and T_2 correspond to total and partial eclipses, respectively. (Adapted from F. H. Shu, *The Physical Universe*.)

If the spectroscopic binary also happens to be an eclipsing binary, such that the distance between the two is much larger than their size, then the angle of inclination is close to $90°$. In this case the masses of the two stars can be determined. Furthermore, the radii of the two stars can also determined in the special case of Algol stars. In such systems, one can clearly identify the beginning and the end of the eclipse. The orientation of stars A and B during a primary eclipse is shown in Figure 13.6. We have assumed that the orbit is a circle of radius a. The corresponding light curve is also shown in Figure 13.6.

From the figure it is clear that

$$\theta = \frac{2R_A}{a},$$

$$\phi = \frac{2R_B - 2R_A}{a}, \tag{13.18}$$

where R_A and R_B are the radii of the two stars, which are assumed to be small compared to a. We also have

$$\frac{\theta}{2\pi} = \frac{T_1}{T},$$

$$\frac{\phi}{2\pi} = \frac{T_2}{T}, \tag{13.19}$$

where T_1 and T_2 are the time intervals corresponding to the total and partial eclipse, shown in Figure 13.6. Using Equations 13.18 and 13.19, we can determine the radii R_A and R_B. Hence a spectroscopic binary, which is also an eclipsing binary, allows determination of the masses and the radii of both stars in certain circumstances.

13.4 Mass Transfer in Binary Systems

In some binary systems, the distance between two stars is very small. In such cases their mutual gravitational pull can strongly affect the properties and evolution of the stars. Consider two stars A and B of masses M_A and M_B revolving about their mutual center of mass with angular velocity $\vec{\omega}$. We assume that the stars are spherical and their orbit circular. We are interested in the force acting on a test particle located at point P, shown in Figure 13.7. The distance between the two stars $r_A + r_B = a$. Let Φ denote the gravitational potential at point P. It is defined as the gravitational potential energy of a particle of unit mass. Hence the gravitational force on a particle of mass m is given by

$$\vec{F} = -m\vec{\nabla}\Phi. \tag{13.20}$$

We work in a coordinate system rotating about C with angular velocity $\vec{\omega}$. In this frame the stars lie along the x-axis. The gravitational potential at P, including the effect due to centrifugal force, is given by

$$\Phi = -\frac{GM_A}{r_1} - \frac{GM_B}{r_2} - \frac{1}{2}(\vec{\omega} \times \vec{r}) \cdot (\vec{\omega} \times \vec{r}). \tag{13.21}$$

In order to calculate the total force, we also need to include the Coriolis force if the particle has non-zero velocity in the rotating frame.

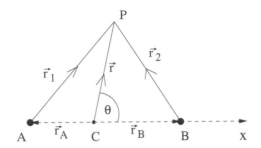

FIGURE 13.7: The test particle P is located at position vector $\vec{r}$ with respect to C, the center of mass of the binary system. Its position vectors relative to stars A and B are $\vec{r}_1$ and $\vec{r}_2$, respectively. The stars A and B are located at positions $\vec{r}_A$ and $\vec{r}_B$ with respect to C. They lie along the x-axis in the rotating coordinate system.

Let us choose coordinates such that $\vec{\omega}$ points in the z direction, that is,

$$\vec{\omega} = \omega\hat{z}. \tag{13.22}$$

We compute the potential, Φ, in the $x - y$ plane. Let $\vec{r}$, lying in $x - y$ plane, make an angle θ with respect to the x-axis. It can be expressed as

$$\vec{r} = r\cos\theta\hat{x} + r\sin\theta\hat{y}. \tag{13.23}$$

We obtain

$$\Phi = -\frac{GM_A}{r_1} - \frac{GM_B}{r_2} - \frac{1}{2}\omega^2 r^2. \tag{13.24}$$

It is convenient to define

$$\Phi_0 = \frac{G(M_A + M_B)}{a}, \tag{13.25}$$

which is the magnitude of the potential at distance a due to a star of mass $M_A + M_B$. In Figure 13.8, we plot the potential, Φ/Φ_0, in the $x - y$ plane. Here we have chosen $M_B/M_A = 0.25$. We notice the two deep potential wells corresponding to the positions of the two stars. These are separated by a saddle point. The name originates due to the resemblance of Φ near this point to the saddle used to ride horses. In the $x - y$ plane, the potential increases as we move away from this point along the negative or positive y-axis. However, the potential decreases along the x-axis. A saddle point corresponds to a position of unstable equilibrium. Hence, a test particle at this point experiences zero force but will move away and fall into one of the potential wells if perturbed slightly.

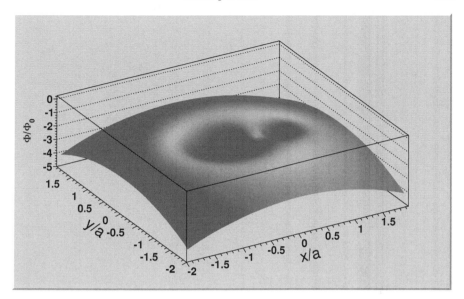

FIGURE 13.8: The gravitational potential, Φ/Φ_0, for two stars of masses M_A and M_B lying along the x-axis, such that $M_B/M_A = 0.25$. The center of mass of the two stars is located at $x = 0$ and $y = 0$.

A contour plot of the potential is shown in Figure 13.9. The saddle point is labeled as L_1 on this plot. It is one of the Lagrange points, L_1, of the binary system. These are points of stable or unstable equilibrium in our rotating coordinate system. At these points, $\vec{F} = -m\vec{\nabla}\Phi = 0$. Consider a particle of small mass that is placed at one of these points and is revolving at angular velocity $\vec{\omega}$ along with the binary system. The particle will continue its motion unless perturbed by some external influence. Hence satellites are often placed at the Lagrange points of the Earth-Sun system. We also show the potential along the x-axis in Figure 13.10. Here we show the location of two more Lagrange points, L_2 and L_3. This also shows clearly the decrease in potential as we move away from L_1 along the x-axis. In Figure 13.11, we show the behavior of the potential near the vicinity of L_1 along the y and z axes. We find that the potential increases as we move away from L_1 in these directions. Hence a particle placed at L_1 will be stable to perturbations in the $y-z$ plane. If the perturbing force has any component in the x direction, the particle will move away from this point.

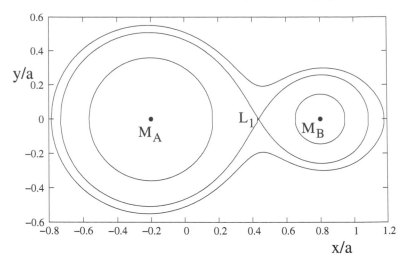

FIGURE 13.9: The contour plot of the gravitational potential, Φ/Φ_0, for the two stars of masses M_A and M_B described in Figure 13.8.

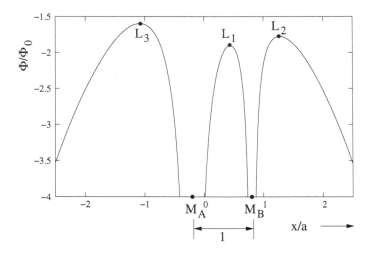

FIGURE 13.10: The gravitational potential, Φ/Φ_0, for the two stars of masses M_A and M_B corresponding to Figure 13.8, as a function of the x coordinate. The distance between the two stars is a. (Adapted from B. W. Carroll and D. A. Ostlie, *An Introduction to Modern Astrophysics*.)

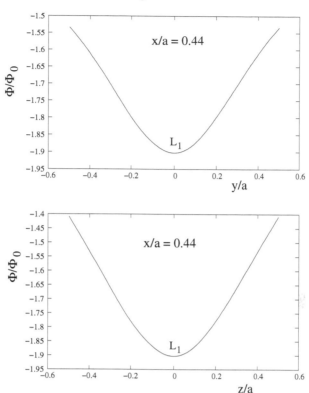

FIGURE 13.11: The gravitational potential, Φ/Φ_0, for two stars of masses M_A and M_B corresponding to Figure 13.8, along the y- and z-axis.

Let us now consider a binary system in which one of the stars starts to leave the main sequence and its size begins to grow. This is shown in Figure 13.12, where we also show the contour that contains the Lagrange point L_1 (see Figure 13.9). The tear-shaped region surrounding each of the stars, lying within this contour, is called the Roche lobe. As star A begins to expand, it eventually fills the Roche lobe. Its shape is no longer spherical due to the influence of the binary partner, B. Once it fills the Roche lobe, material from A starts to flow toward B through the Lagrange point. This matter spirals around B, forming a disk-shaped region called the accretion disk and eventually falls onto the surface of this star. Due to the transfer of mass from star A to B, the evolution of both stars is affected significantly. This explains, for example, why the more massive star, Algol A, in the Algol system is still on main sequence, whereas Algol B is in the giant phase. As mentioned earlier, this can happen if the distance between the two binary partners is relatively small. Such systems are identifiable by their short time period.

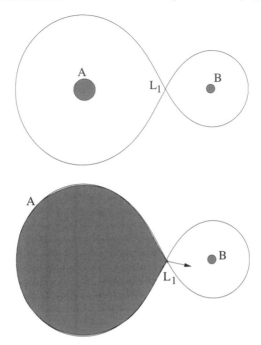

FIGURE 13.12: These figures illustrate the mass transfer from star A to its binary partner B. As star A begins to expand, it fills up its Roche lobe, as shown in the lower figure. Material from star A starts to flow toward B at the Lagrange point L_1.

Exercises

13.1 Assume that two stars of the same mass are moving in a circular orbit around one another. Make a plot analogous to Figure 13.1. Repeat this for the case where one is twice as heavy as the other.

13.2 Consider a binary star system of eccentricity 0.8. Make a rough sketch of the trajectories of the two stars relative to their center of mass.

13.3 For single-lined spectroscopic binaries, we can find the speed v_{Ar} of only one of the stars, besides the time period. Show that this allows determination of only the combination

$$\frac{(m_B \sin \eta)^3}{(m_A + m_B)^2} = \frac{v_A^3 T}{2\pi G}.$$

13.4 In this problem we will study the properties of a binary system whose orbit is tilted relative to the line of sight. For simplicity we assume that the true orbit is circular. We choose one of the stars as the origin O of our coordinate system. The binary partner moves in a circular orbit of radius a about O. As explained in text, the observed orbit is an ellipse with one of the stars at its center. The normal to the orbital plane, $\hat{n}$, makes an angle η with respect to the line of sight, as shown in Figure 13.13. This is called the tilt angle. We assume that $\hat{n}$ lies in the $x - y$ plane and the line of sight is along the x-axis. In that case,

$$\hat{n} = \cos\eta\,\hat{x} + \sin\eta\,\hat{y}\,.$$

The position vector of the star in orbit with respect to the origin is

$$\vec{r} = x\hat{x} + y\hat{y} + z\hat{z}\,.$$

(a) Show that the condition, $\hat{n} \cdot \vec{r} = 0$, leads to

$$x = -y\tan\eta\,.$$

(b) Let a be the radius of the circle. Show that setting $r^2 = a^2$ leads to

$$\frac{y^2}{\cos^2\eta} + z^2 = a^2\,.$$

This is the equation of the observed trajectory, that is, the true trajectory projected onto the $y - z$ plane.

(c) Let v and ω represent the linear and angular speeds of the star. We can set $z = a\cos(\omega t)$. Using the equations derived above, determine the time dependence of y and x.

(d) Determine the velocity vector

$$\vec{v} = \frac{d\vec{r}}{dt}\,.$$

Verify that $v^2 = (a\omega)^2$.

(e) By Doppler shift we can measure v_x, the x-component of the velocity vector. Using the result in (d), verify that the maximum value of v_x is $v\sin\eta$.

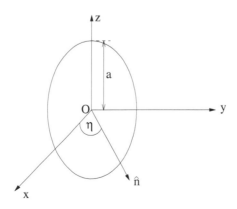

FIGURE 13.13: A tilted circular orbit of a binary system. The normal to the orbital plane is denoted as $\hat{n}$ and the line of sight is along the x-axis.

Chapter 14

The Milky Way

Stars are not found in isolation but in huge clusters called galaxies. The Sun is part of the Milky Way, which is a barred spiral galaxy and contains more than 200 billion stars. The main features of the Milky Way are the central bulge, the disk, and the halo, as shown in Figure 1.2. The disk diameter is about 30 Kpc and disk thickness approximately equal to 1 Kpc. Surrounding the disk is the galactic halo, which extends much beyond the disk. The Sun is located at about 8.5 Kpc from the center. The density of stars is much higher in the disk in comparison to the halo. It is also very high toward the center of the galaxy, which is located in the constellation Sagittarius. Away from the city lights, one can observe the disk as a white band stretching across the entire sky. Nearly all the objects that are directly visible in the night sky lie within the Milky Way. The only exception is the Andromeda galaxy, the closest bright spiral galaxy, and the Large and Small Magellanic Clouds, which are irregular dwarf galaxies and can be seen from the southern hemisphere.

Historically the first detailed study to map the Milky Way was done by William Herschel. In order to determine the distribution of stars in three dimensions, one needs to know their distance. Measurement of astronomical distances is difficult. We will discuss the detailed process by which astronomers have been able to map the Universe in the next section. Herschel side-stepped this problem by making some simplifying assumptions. He essentially assumed that if the number of stars in a particular direction is small, then the edge of the Milky Way is close to us in that direction. By this procedure, Herschel, in 1785, concluded that the Milky Way is a disk-like structure with the Sun close to its center.

We now know that the Sun is very far from the center of the Milky Way. Herschel came to the wrong conclusion because the real center is obscured from our view by dust. Hence we are unable to directly observe the true density of stars in this direction. The true extent of the Milky Way was first determined by Harlow Shapley by mapping the globular clusters between the years 1914 and 1919. These clusters are found predominantly in the halo, which is relatively free of dust. He was able to determine the distance of these objects and deduced that they form a spherically symmetric distribution, whose center is located very far away from the Sun. He concluded that this must be the true center of the Milky Way, which must be much larger than what was previously believed.

14.1   The Distance Ladder

The determination of astronomical distances played a crucial role in the discovery of the true extent of the Milky Way. In this section we give a detailed description of how these distances are estimated.

The distance of nearby stars can be measured by the parallax technique, discussed in Section 3.3. For modern ground-based telescopes, this is useful only for distances of about 100 parsec. For the measurement of larger distances, astronomers try to deduce the luminosity of an astronomical object. Its distance can then be determined by the measurement of its flux. The application of this procedure to measure distances of star clusters was discussed in Section 6.8. In order to use this technique we need to calibrate the relationship between luminosity and flux or, equivalently, the absolute and apparent magnitude of a cluster. For this we require one cluster whose distance can be measured by some other technique. The nearest star cluster is an open cluster called Hyades, located at a distance of 47 parsec. Hence a lot of effort was devoted to a measurement of its distance during the twentieth century. At early times, ground-based telescopes were unable to measure its annual parallax reliably. Hence astronomers had to invent an alternate technique.

Historically the distance of the Hyades cluster was determined by the moving cluster method. It is assumed that all stars within a cluster have the same space velocity. This is reasonable since they are gravitationally bound to one another. We can measure the radial velocities, V_r, of several stars using the Doppler effect. We can also measure their proper motion μ. The transverse component of their velocity $V_t = \mu d$, where d is the distance of the cluster. Let us choose coordinates such that the center of the cluster lies on the z-axis. For simplicity, let us assume that the velocities of all the stars in this cluster point along the z-axis. These are shown schematically in Figure 14.1 by vectors parallel to the z-axis. Consider a particular star located at an angle θ to the z-axis. The transverse component of the velocity of this star points toward the point C. Similarly, the transverse components of all stars point toward C, called the point of convergence, as shown in Figure 14.2. Using the standard relation between the two components of a vector,

$$\tan \theta = \frac{V_t}{V_r}, \tag{14.1}$$

and $V_t = \mu d$, we obtain

$$d = \frac{V_r \tan \theta}{\mu}. \qquad (14.2)$$

A measurement of V_r, μ, and θ therefore yields an estimate of the distance.

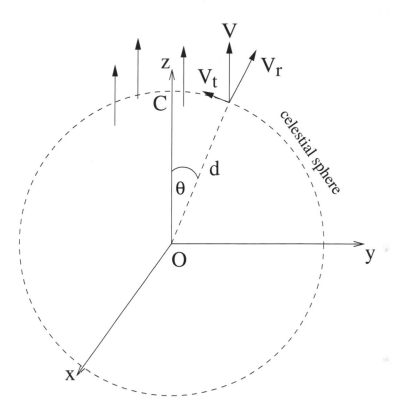

FIGURE 14.1: An illustration of the velocity vectors of stars within a cluster. These vectors projected onto the celestial sphere converge to a point C.

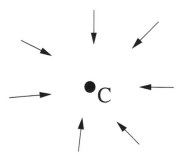

FIGURE 14.2: The velocity vectors of stars within a cluster, projected onto the celestial sphere, appear to converge to a point.

In order to get a reliable estimate, it is necessary to measure these observables for several stars and obtain the best fit value of d. Here we have assumed that the motion of the cluster is parallel to the line of sight. This is not true in general. However, the principle and the resulting formula, Equation 14.2 is valid in general. The reader can prove that as an exercise. Note that, in the special case discussed above, the point of convergence lies within the cluster. In general, this is not true.

Once the distance to the Hyades cluster was determined, it was used to calibrate the relationship between color and the absolute magnitudes of the main sequence stars within an open cluster. This can be used to determine the distance to other clusters, as discussed in Section 6.8. This method can be used to measure much larger distances. However, it could not be used to determine the distance of globular clusters, which was required to determine the true extent of the Milky Way. The problem is that globular clusters are relatively old, and the stars in these clusters that are still on the main sequence have relatively low mass. These main sequence stars are not very bright and cannot be observed at such large distances.

In order to determine larger distances, astronomers use variable stars, such as Cepheids and RR Lyrae variables. Cepheid variables are giant or super-giant stars and show periodic variation of luminosity. Their periods typically range from a day to about 100 days. We can measure their mean apparent magnitude and period. The important point is that their mean luminosities or absolute magnitudes, M_V, are related to their period, T. The Period-Luminosity (P-L) relationship can be expressed as

$$M_V = -2.81 \log(T) - 1.43, \tag{14.3}$$

where the period is expressed in days. Because the period can be directly measured, their absolute magnitudes can be determined. Such objects, whose absolute magnitudes can be deduced, are called standard candles. Using the inverse square law, which in terms of magnitudes is given by Equation 4.24, we can determine their distance. The RR Lyrae variables have properties similar to Cepheids, but have smaller time period. In 1914, when Shapley attempted to map the Milky Way, the period-luminosity relationship of Cepheids had not been calibrated. Earlier, Henrietta Swan Leavitt had deduced the correlation between period and luminosity by observations of 16 Cepheids from Small Magellanic Cloud (SMC). However, she did not know the distance of SMC. Hence she was unable to provide an absolute calibration. Due to their relatively low abundance, they are found only at large distances. Hence, it was not possible to measure the distance to any one of them by annual parallax. Shapley solved this problem using a statistical technique to measure the distance of 11 Cepheid variables whose proper motions were known. The basic idea is that a star at larger distance is expected to have smaller proper motion. Hence one can assume an inverse relationship between these two observables and deduce the proportionality constant by making a fit to data. Using this procedure, he was able to calibrate the P-L relationship of Cepheids.

There were actually several errors in Shapley's observations. In particular, he neglected interstellar absorption. Furthermore, he misidentified W Virginis stars in globular clusters as Cepheid variables and applied his P-L relationship to these in order to determine the distance of these clusters. Fortunately, the different errors tended to cancel one another and his estimates of the distances to globular clusters were approximately correct. Hence he was able to correctly infer the true size of the Milky Way.

The P-L relationship of Cepheid variables also proved useful for finding objects beyond the Milky Way. Using this technique, Edwin Hubble, in 1923, estimated that the distance of the nebula M31 was 280 Kpc. Due to the errors in calibration of Cepheid P-L relationship, this was underestimated by a factor of about 3. This nebula is the next nearest prominent galaxy, called Andromeda. Its distance is now known to be 778 ± 33 Kpc.

The Cepheid variables are not bright enough to go much beyond a few Mpc. In order to go to larger distances, one needs brighter standard candles. One possibility is to use the brightest stars in a galaxy. Statistically, their brightness appears to be same in all the nearby galaxies and hence these can act as standard candles. This method is feasible for distances up to about 10 Mpc. Another possibility is to use the luminosities of globular clusters. This requires observation of the magnitudes of many globular clusters within a galaxy. These are used to determine the Globular Cluster Luminosity Function (GCLF), $\phi(M)$, which is the probability to find a globular cluster with absolute magnitude M. For blue magnitude M_B, this shows a peak at $M_B \approx -6.5$. For a distant galaxy we can determine the distribution of apparent magnitudes, m_B, of globular clusters. This can be used to determine its distance because the absolute magnitude of the peak position is known. This method can be used to measure distances less than 100 Mpc.

Another useful method is the Tully–Fisher relationship for spiral galaxies. This relates the luminosity of a galaxy to its rotational velocity and will be discussed in more detail in the next chapter. The luminosity of a distant galaxy can, therefore, be deduced from the observation of its rotational speed. This method can be used for distances larger than 100 Mpc.

For the largest distances, the most effective method has been the light curve of Type Ia supernovae. These originate in a binary star system in which one of the stars is a white dwarf. This star accretes mass from its binary partner. Once its mass exceeds the Chandrasekhar limit (1.4 solar masses), it undergoes a supernova explosion. Recall that this is the maximum possible value of the mass of a white dwarf. Beyond this value it becomes unstable. The supernova luminosity rises quickly and reaches its peak value within a few days. This is followed by a slow decline. A typical light curve for such a supernova is shown in Figure 14.3. The important point is that the peak luminosity of all Type Ia supernovae is approximately the same. This is understandable because, in all cases, the mass of the white dwarf that explodes is the same. Hence the nature of the supernova explosion is the same in all cases. The peak absolute visual magnitudes of all such supernovae is approximately −19.3. There is a small spread in these values. It turns out that the peak value is related to the width of the light curve. Using this relationship and the measured value of the width, the peak value can be fixed with higher accuracy. Hence these supernovae act as standard candles and their distance can be determined.

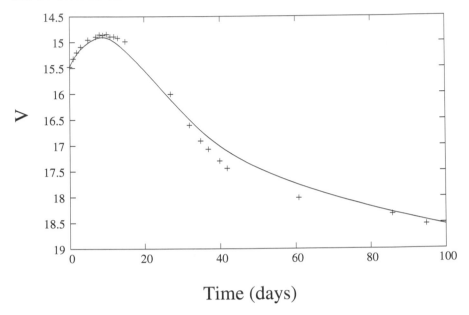

FIGURE 14.3: The light curve of the Type Ia supernova SN1999dq. The *x*-axis shows the time in days and the *y*-axis the apparent *V* magnitude.

14.2   Distribution of Matter in the Milky Way

The main components of the Milky Way are the thin disk, the halo, and the central bulge, as illustrated in Figure 1.2. The density of matter is much higher in the disk in comparison to the halo. The disk contains stars that are relatively young compared to those in the halo. It also contains a considerable amount of gas and dust and many regions of active star formation. In contrast, the halo contains negligible gas and dust and hence no star formation.

The structure of the disk depends on the observed component. The gas and dust lie within a region that has a thickness of about 100 pc. New stars take birth in this region. Hence it also contains a large number of OB associations. If we map the distribution of low mass main sequence stars, such as the Sun, the disk appears to have a thickness of about 1 Kpc.

Annotated Roadmap to the Milky Way
(artist's concept)

NASA / JPL-Caltech / R. Hurt (SSC-Caltech)   ssc2008-10b

FIGURE 14.4: An illustration of the known spiral arms of the Milky Way galaxy. (Image courtesy of NASA/JPL-Caltech.)

A very interesting feature of the disk is the spiral arms. If we observe the OB associations close to the Sun, we find that they are not distributed uniformly within the disk but rather are concentrated in regions that are identified as parts of the spiral arms of our galaxy. The Sun itself lies on the inner edge of the Orion–Cygnus arm. We observe the Carina–Sagittarius Arm toward the center of the galaxy and the Perseus arm in the opposite direction. We cannot map the disk at large distances at visual frequencies due to the presence of the intervening dust. At larger distances, the structure of the disk can be determined by radio observations of the 21 cm spectral line emitted by neutral hydrogen. The gas also undergoes rotation around the galactic center. Hence the gas at different distances has different radial velocities. This velocity component can be determined by measuring the Doppler shift of the 21 cm line. These observations allow one to map the entire disk and obtain the full extent of the spiral arms, as explained below. An illustration of the different arms is shown in Figure 14.4. We point out that the detailed spiral structure of the galaxy is not well known and different astronomers do not agree on the precise nature of these arms.

There is considerable evidence that suggests the presence of a supermassive black hole, with mass about a million times the mass of the Sun, at the center of the Milky Way. This region is called Sagittarius A*. The mass is deduced from measurements of the stellar velocities in this region. Observations suggest that the density of this region is sufficiently high that it would collapse to form a black hole during the lifetime of the Milky Way. Besides the measurement of mass, other observations such as radio emission are also consistent with the presence of a black hole.

The luminosity distribution of stars in the Milky Way is characterized by the luminosity function, $\phi(L)$, where L is the stellar luminosity. Consider a certain region of the Milky Way whose volume is V. We count the number of stars in this volume whose luminosities lie between L and $L + dL$. Dividing this number by V and dL gives us the luminosity function, $\phi(L)$. It is equal to the number of stars per unit luminosity, per unit volume. It is convenient to define this function in terms of the absolute magnitude M. Hence $\phi(M)dM$ is the number density of stars with absolute magnitudes lying between M and $M + dM$. Observationally, it is determined by counting the number of stars that have absolute magnitudes between $M - 1/2$ and $M + 1/2$ in the region of interest and dividing the resulting number by the total volume. This function can be determined reliably only in the neighborhood of the Sun.

The luminosity function is important for studying the structure of the Milky Way. In general, it may have a complicated dependence on position. However, as a first approximation, it may be reasonable to assume that it is constant within a particular component of the galaxy. For example, it may be uniform within the disk or the halo. Hence its measurement in our neighborhood may provide a good approximation of its value within the disk.

Let $n(\vec{r})$ be the number density of stars at position $\vec{r}$. Let us suppose that we are interested in obtaining it in the galactic disk. Let $dN(m, l, b)$ be the observed number of stars in the direction (l, b) within the solid angle $\Delta\Omega$ and the radial interval $(r, r + dr)$. Their apparent magnitudes lie in the interval $(m - 1/2, m + 1/2)$. We can express $dN(m, l, b)$ in terms of the luminosity function, by the equation

$$dN(m, l, b) = n(\vec{r})\frac{\phi(M)}{n_0(\vec{r})}d^3r \,, \tag{14.4}$$

where $n_0(\vec{r})$ is the number density in the neighborhood of the Sun and $d^3r = r^2 dr \Delta\Omega$. We need to divide by n_0 because $\phi(M)$ is the number of stars per unit volume in our neighborhood. We define the relative number density $D(\vec{r})$ by the relationship

$$D(\vec{r}) = \frac{n(\vec{r})}{n_0(\vec{r})} \,. \tag{14.5}$$

We can express the absolute magnitude of a star in terms of its apparent magnitude by using Equation 7.18. Hence the number of stars $N(m, l, b)$ in the direction (l, b) with apparent magnitude lying in the range $(m - 1/2, m + 1/2)$ can be expressed as

$$N(m, l, b) = \int_0^\infty D(\vec{r}) \phi \left(m - 5 \log \frac{r}{10 \text{ pc}} - A \right) r^2 dr \Delta\Omega. \tag{14.6}$$

This equation, first obtained by von Seeliger, is called the fundamental equation of stellar statistics. It allows one to deduce the number density of stars in any region of the Milky Way by observing $N(m, l, b)$ and the luminosity function. As explained earlier, we can determine the luminosity function of the stars in our neighborhood, which provides a reasonable approximation of its value within the disk.

14.3 Differential Rotation of the Milky Way

The stars, along with the gas and dust in the Milky Way disk, rotate about the center of the galaxy in nearly circular orbits. The rotation is differential, that is, the disk does not rotate as a rigid body. The angular speed depends on the distance from the center. Before studying this in detail, we point out that the motion of stars in the galactic halo is very different from that in the disk. In general, the halo stars display elliptical orbits with high eccentricity.

The differential rotation of the disk is not observable if we consider stars in our neighborhood, that is, located at distances less than 100 pc from us. These stars appear to be moving in random directions with speeds ranging from a few Km/sec to more than 100 Km/sec. It is convenient to define a reference frame, called the Local Standard of Rest (LSR), in which the mean velocity of all stars in our neighborhood, including the Sun, is zero. The velocity of a star with respect to the LSR is called its peculiar velocity. We can divide the entire disk into similar domains of area roughly equal to $(100 \text{ pc})^2$ and determine the mean velocity vector of stars within each domain. It is this mean motion of different domains that shows circular motion around the galactic center.

Consider a star P in the galactic disk at distance R from the center, located at galactic latitude l, as shown in Figure 14.5. Let us assume that it rotates about the center in a circular orbit with angular speed $\omega(R)$. We next determine its radial and tangential velocities, as observed from the Sun. Let us denote these by symbols v_r and v_t, respectively. Let the linear velocity of the star and the Sun be $\vec{V}$ and $\vec{V}_S$, respectively, relative to the galactic center. We clarify that $\vec{V}_S$ represents the velocity of the Sun after subtracting out its peculiar motion relative to the LSR. Hence it really is the velocity of the LSR relative to the galactic center. Similarly, $\vec{V}$ is the mean velocity vector of another group (or domain) of stars located at mean distance R from the galactic center. The magnitudes of these velocity vectors are given by

$$\begin{aligned} V &= \omega R, \\ V_S &= \omega_S R_S, \end{aligned} \tag{14.7}$$

where ω_S is the angular speed of the Sun and R_S its distance from the center. The distance of the star P from the Sun along the line of sight is r. The velocity of the star with respect to the Sun is

$$\vec{v} = \vec{V} - \vec{V}_S. \tag{14.8}$$

Let θ be the angle between the line of sight and the radius R, as shown in Figure 14.5. The radial and tangential speeds of the star, with respect to the Sun, are

$$\begin{aligned} v_r &= V \sin(\theta) - V_S \sin(l), \\ v_t &= V \cos(\theta) - V_S \cos(l), \end{aligned} \tag{14.9}$$

respectively. We see from Figure 14.5 that

$$R \sin(\theta) = R_S \sin(l), \tag{14.10}$$

$$r = R_S \cos(l) - R \cos(\theta). \tag{14.11}$$

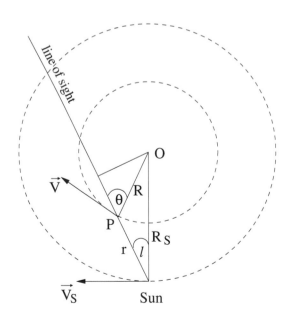

FIGURE 14.5: A star P at distance R from the center rotates around the galactic center O with angular speed ω. Its linear velocity is $\vec{V}$.

Hence, using Equations 14.9 and 14.7, we obtain

$$
\begin{aligned}
v_r &= (\omega - \omega_S)R_S \sin(l)\,, \\
v_t &= (\omega - \omega_S)R_S \cos(l) - \omega r\,.
\end{aligned}
\tag{14.12}
$$

Assuming that the star P is not very far away from the Sun, we expect that $\omega \approx \omega_S$. Hence we can expand $\omega(R)$ by making a Taylor expansion. We obtain

$$
\omega(R) = \omega_S + (R - R_S)\frac{d\omega}{dR}\bigg|_{R=R_S}\,.
\tag{14.13}
$$

Furthermore, $R \approx R_S \gg r$. This implies, using Equations 14.10 and 14.11, that

$$
R_S \approx R + r\cos(l)\,.
\tag{14.14}
$$

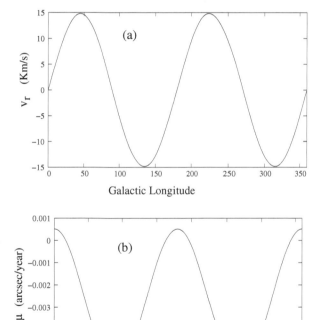

FIGURE 14.6: The radial speed v_r and proper motion μ of stars in the galactic disk, as a function of the galactic latitude, due to their differential rotation about the center of the Milky Way. These curves are obtained using the measured values of the Oort constants.

Hence we obtain

$$v_r = Ar\sin(2l),$$
$$v_t = Ar\cos(2l) + Br, \tag{14.15}$$

where A and B are Oort constants, defined as

$$A = -\frac{R_S}{2}\frac{d\omega}{dR}\bigg|_{R=R_S},$$

$$B = -\left(\omega_S + \frac{R_S}{2}\frac{d\omega}{dR}\bigg|_{R=R_S}\right). \tag{14.16}$$

These constants contain information about the motion of stars in the neighborhood of the Sun. The constant A is directly proportional to the rate of change of ω with R. It would be zero if the galaxy rotates as a rigid body. The constant $(A - B)$ is equal to ω_S, the angular speed of the Sun or, more precisely, the local standard of rest (LSR) about the galactic center. These constants can be determined by observing the radial and proper motions of stars. The proper motion, $\mu = v_t/r$, is given by

$$\mu = A\cos(2l) + B. \tag{14.17}$$

The observed values of these constants, using the Hipparcos satellite data, are

$$A = 14.82 \pm 0.84 \text{ Km s}^{-1} \text{ Kpc}^{-1},$$
$$B = -12.37 \pm 0.64 \text{ Km s}^{-1} \text{ Kpc}^{-1}. \tag{14.18}$$

The fact that $A \neq 0$ directly implies that the galaxy does not rotate as a rigid body. We also find that $\omega_S = A - B = 27.19$ Km/(s·Kpc). The Sun is at a distance of about 8.3 Kpc from the galactic center. Hence this leads to a linear speed of Sun, $V_S \approx 226$ Km/s. In units of radians per year, $\omega_S = 2.8 \times 10^{-8}$. This implies that the time period of rotation of the Sun around the galactic center is approximately 2.3×10^8 years. In Figure 14.6, we plot the radial speeds and proper motions of stars as a function of the galactic longitude using these observed values.

14.4   Mapping the Galactic Disk with Radio Waves

As we have mentioned above, radio waves can travel large distances within the Milky Way and hence can be used to map the galactic disk. The mapping is done mainly by using the 21 cm spectral line that is emitted by atomic hydrogen, due to its hyperfine splitting.

The basic idea is explained in Figure 14.7. If we look in any direction, we are likely to see several different hydrogen clouds located at different distances. For illustration, we show four different clouds A, B, C, and D along the line of sight, at the galactic longitude l, on the galactic disk. The cloud B is closest to the galactic center O. We assume that the angular speed (ω) of the galaxy

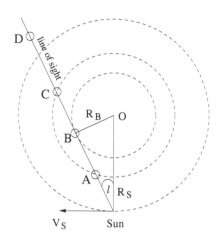

FIGURE 14.7: An illustration of a series of hydrogen clouds A, B, C, and D lying along the line of sight, at galactic longitude l, on the galactic disk. These clouds rotate about the galactic center with speeds given by $\omega(R)$, where R is the distance from the center. The cloud B is closest to the galactic center and has maximum value of ω. This leads to the largest radial speed, v_r, relative to the Sun. (Adapted from H. Karttunen et al., *Fundamental Astronomy*.)

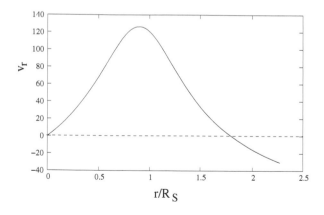

FIGURE 14.8: The radial speed v_r of a hydrogen cloud as a function of distance (r) from the Sun at fixed galactic longitude l for the range $0 < l < 90°$.

depends only on R, that is,

$$\omega = \omega(R), \tag{14.19}$$

where R is the distance from the center. Observations show that $\omega(R)$ decreases with R. A rough approximation to this relationship is obtained by noting that the linear speed of rotation is approximately constant beyond a few Kpc from the center. Hence at such distances, we can assume that $\omega(R) = V(R)/R \approx V_0/R$, where V_0 is a constant. This is of course a very rough approximation and is being used here only for illustration. Using Equation 14.12, for $0 < l < 90°$, we see that at fixed l, as we move away from the Sun, v_r first increases and then starts to decrease with r. Here r is the distance of the cloud from the Sun. This dependence of v_r on r is illustrated in Figure 14.8. The maximum value of radial speed corresponds to the hydrogen cloud located at B in Figure 14.7. For $90° < l < 270°$, the line of sight will only encounter clouds at distances larger than R_S from the galactic center, as can be seen from Figure 14.7. For such clouds, $\omega(R) < \omega_S$. In this case, Equation 14.12 implies that $v_r < 0$ and continues to decrease monotonically with r. These clouds appear to move toward the observer. For the range $270° < l < 360°$ (or equivalently $-90° < l < 0o$), the radial speed v_r first decreases, that is, becomes negative, reaches a minimum, and then starts to increase with r. This behavior is exactly opposite (or inverted) in comparison to that seen in Figure 14.8.

The important point is that if we can identify the cloud with the maximum speed, then we know that it is located at the minimum distance from the Sun. By geometry, we deduce that its distance $R_B = R_S \sin(l)$. The speed of the cloud can be deduced by the Doppler shift of its 21 cm spectral line. Recall

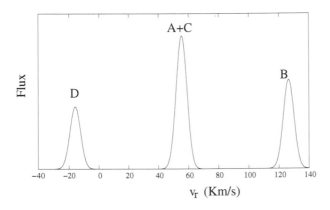

FIGURE 14.9: An illustration of the line profile due to the radiation received from clouds A, B, C and D along the line of sight at fixed galactic longitude l. On the x-axis we show the radial speed of the cloud, which has been deduced from the observed Doppler shifted wavelength of the 21 cm hydrogen line, as explained in text.

that if the source is moving away from us with speed, v_r, then the shift in the wavelength of its spectral line is $\Delta\lambda/\lambda_0 = v_r/c$, where λ_0 is the wavelength observed by an observer at rest relative to the source. Hence if a cloud is moving away from us, then we expect to see a Lorentzian intensity profile with the peak intensity at the wavelength $\lambda = \lambda_0(1 + v_r/c)$. If we have several different clouds at different distances along the line of sight, then the spectrum will be a sum of the line profiles due to individual clouds. This is schematically illustrated in Figure 14.9, where we show the observed spectrum assuming only four clouds, A, B, C, D. On the x-axis we show v_r as deduced from the relationship $\lambda = \lambda_0(1 + v_r/c)$. By identifying the cloud with the largest Doppler shift and hence largest v_r, we can deduce the angular speed, ω, at the distance R_B. The angular speed can be determined by using Equation 14.12. This procedure repeated for all l allows us to determine $\omega(R)$, that is, ω as a function of distance from the galactic center. Once this is determined, we can deduce the distance of different clouds from the galactic center by their observed values of v_r. Their observed flux then directly leads to the intensity profile of the galactic disk.

14.5 Formation of the Spiral Arms

The spiral arms are regions of higher density of matter in the disk of spiral galaxies, such as the Milky Way. Here we briefly discuss how these structures are formed and how they remain preserved over many cycles of rotation of these galaxies. One might naively expect that the spiral structure rotates at the same rate as the stars. However, if this were true, then, due to differential rotation, the spiral will wind around the center in a few complete cycles of rotation of the galaxy. This is illustrated in Figure 14.10. A more likely possibility is that the individual stars move at higher speeds compared with the motion of the spiral pattern. This may be understood by an analogy, described in *The Physical Universe, An introduction to Astronomy* by Frank H. Shu. Consider a highway with two lanes, such that one of the lanes is broken down at some point. As cars approach this point, they will slow down, slowly change lanes, and move on. Furthermore, one will observe a small clustering of cars in the vicinity of this point. This cluster can be compared with the spiral pattern. In this case, the cluster does not move at all, while the individual cars keep moving in and out of this region. In a similar manner, a star slows down while crossing a spiral arm. Eventually it completes its motion through the arm and moves on. Hence the spiral arm is composed of different stars at different times but its shape remains similar. The spiral pattern rotates at a much slower rate compared with individual stars.

The formation of spiral patterns is best described by the density wave theory, originally proposed by Lin and Shu in 1963. The basic idea is that the disk does not have exact axial symmetry. For example, the density may be a little higher at some azimuthal angles ϕ compared to the mean density. Hence, the force acting on a star located on the disk depends also on the azimuthal angle ϕ, besides its dependence on the radius r. We may assume that initially the ϕ-dependent part of the force is a small perturbation. Depending on the nature of this perturbation, the azimuthal asymmetry may be enhanced with time. This means that the density contrast may start to increase, leading to a further increase in the ϕ-dependent perturbating force and so on. This process may eventually lead to the formation of spiral arms.

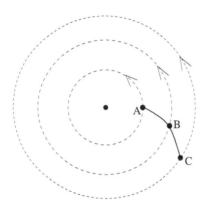

FIGURE 14.10: Consider three stars, A, B, and C lying on a spiral arm. Let us assume that star A moves at angular speed twice that of C due to differential rotation of the galaxy. Hence in a time interval taken by A to complete two rotation cycles, C would have completed just one. Hence the spiral pattern will form a complete loop that will be further distorted with more cycles of rotation. The observed spirals do not show such behavior. This indicates that the spiral pattern itself does not rotate with the same speed as the individual stars.

Exercises

14.1 Consider a cluster of stars located along the z-axis. The velocity vectors of stars in three-dimensional space are shown in Figure 14.11. In general, the point of convergence of these velocity vectors, projected onto the celestial sphere, lies outside the cluster. This point is labeled C in Figure 14.11. Show that the distance to the cluster is given by the Equation 14.11, even in this general case.

14.2 Prove the approximate Equation 14.14, by using Equation 14.10. Using this, obtain the formulae for v_r and v_t given in Equation 14.15. You can use

$$\left.\frac{d\omega}{dR}\right|_{R=R_S} = \frac{1}{R_S}\left.\frac{dV}{dR}\right|_{R=R_S} - \frac{V_S}{R_S^2}, \tag{14.20}$$

$$A = \frac{1}{2} \left(\omega_S - \frac{dV}{dR}\bigg|_{R=R_S} \right),$$

$$B = -\frac{1}{2} \left(\omega_S + \frac{dV}{dR}\bigg|_{R=R_S} \right). \tag{14.21}$$

14.3 Using the values of the Oort constants, Equation 14.18, determine the angular speed of the Sun.

14.4 Determine the Oort constants assuming that (a) the Milky Way rotates as a rigid disk, and (b) the angular speed of stars about the galactic center follows Kepler's law.

14.5 Supernovae Type Ia play a very useful role in cosmology because they are standard candles and can be observed at very large distances due to their brightness. Their peak absolute visual magnitude $M_V \approx -19.3$. The limiting apparent magnitude that the Hubble space telescope can detect is $V = 31$. (a) Ignoring interstellar extinction, find the maximum distance over which it can observe such a supernova. Note that this calculation will produce an overestimate because we have ignored extinction and also because the peak will be shifted to lower frequencies due to the cosmological redshift. (b) Repeat this exercise for a typical galaxy such as the Milky Way, which has an absolute magnitude of -20.5.

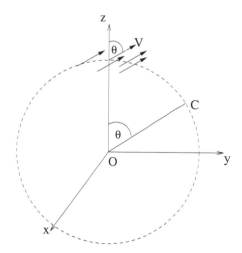

FIGURE 14.11: A schematic illustration of velocity vectors of stars within a cluster. These vectors projected onto the celestial sphere converge to a point C.

Chapter 15

Galaxies

Galaxies are gravitationally bound systems that consist of a huge number of stars. The number of stars in a galaxy vary over a wide range, roughly 10^7 to 10^{14}. For example, the Milky Way galaxy contains more than 200 billion stars. There are at least 100 billion galaxies in the Universe. The galaxies themselves group together to form larger structures called groups or clusters of galaxies. The largest structures seen in the Universe are called superclusters, which contain a large number of clusters. The scale of the largest supercluster is found to be about 100 Mpc. For comparison, the size of the observable Universe is a few Gpc.

The galaxies occur in a wide range of shapes and sizes. They are broadly divided into four different categories called elliptical, lenticular, normal spirals, and barred spirals. Each of these is further subdivided into different types. For example, the elliptical galaxies are classified into different classes labeled E0, E1, E2, etc., depending on their eccentricity. Let the major and minor axes of an elliptical galaxy be a and b, respectively. It is classified as En, where,

$$n = 10(1 - b/a).$$

Hence an E0 galaxy appears circular whereas an E7 would be highly eccentric. These different classes can be depicted in a tuning fork diagram, Figure 15.1, originally due to Hubble.

In Figure 15.2, we show the image of the elliptical galaxy NGC 1132. An image of a normal spiral and barred spiral galaxy is shown in Figures 15.3 and 15.4, respectively. A barred spiral, such as the Milky Way, displays a bar-like structure, as shown in Figure 15.4. The spiral galaxies are classified into Sa, Sb, Sc, depending on the nature of the spiral arms and the size of the central bulge. A spiral galaxy of type Sa has broad, tightly wound spiral arms and a large central bulge. In contrast, a type Sc galaxy has a smaller central bulge and an open spiral pattern. Its spiral arms are narrower and well defined in comparison to a galaxy of type Sa. Similarly, we have a division of barred spirals into SBa, SBb, SBc, etc. The lenticular galaxies, SO or SB0, have structures intermediate between elliptical and spiral.

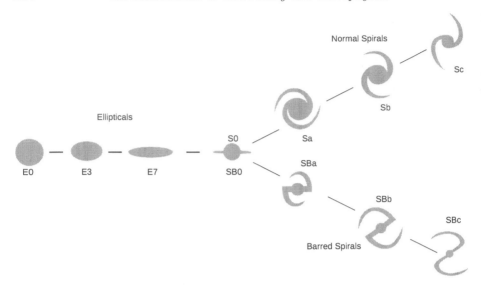

FIGURE 15.1: The Hubble classification of galaxies. The intermediate galaxies, SO and SBO, are called lenticular.

FIGURE 15.2: The elliptical galaxy NGC 1132 observed by the Hubble telescope. (Image courtesy of NASA, ESA, and the Hubble Heritage (STScI/AURA)-ESA/Hubble Collaboration.)

FIGURE 15.3: The spiral galaxy M100 observed by the Hubble telescope. (Image courtesy of NASA, STScI.)

FIGURE 15.4: The barred spiral galaxy NGC 1300 observed by the Hubble telescope. (Image courtesy of NASA, ESA, and The Hubble Heritage Team (STScI/AURA).)

The Hubble classification is purely empirical. However, it was earlier thought to represent the different stages of evolution of galaxies. It was believed that galaxies evolved from elliptical to normal spirals or barred spirals. Hence, elliptical galaxies were designated as early type and spirals and barred spirals as late types. This interpretation is false; however, this terminology is still used occasionally.

Besides these galaxies, we also see other types referred to as irregular galaxies and dwarf galaxies. As the name suggests, an irregular galaxy does not exhibit a simple geometric shape. It need not have some of the structures associated with other galaxies, such as a central bulge. These galaxies are most likely formed out of collisions of two galaxies. A dwarf galaxy is much smaller in size compared to a spiral or an elliptical galaxy. Some prominent examples are the Large Magellanic Cloud and the Small Magellanic Cloud in our astrophysical neighborhood, located, respectively, at a distance of 50 Kpc and 61 Kpc from Earth.

15.1   Elliptical Galaxies

Elliptical galaxies show a relatively simple structure. A typical example is shown in Figure 15.2. They appear as elliptical structures with a monotonic decrease in the density of stars as we move outward from the center. In most cases, they contain negligible interstellar matter, such as gas or dust, and hence there is no new star formation. The stars are very old and formed at the same time. Hence these galaxies are dominated by low mass stars and display a yellow-red color. These galaxies are more likely to be found near the centers of galactic clusters. Within our astrophysical neighborhood, roughly 10 to 15 % of the galaxies are elliptical.

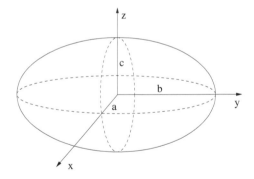

FIGURE 15.5: An ellipsoidal with the three principal axes of length $2a$, $2b$, and $2c$.

Their appearance is determined by their actual three dimensional shape as well as their angle of projection on the sky. Their true shape is roughly ellipsoidal, illustrated in Figure 15.5. In a Cartesian coordinate system, with the origin located at the center of the galaxy, the three-dimensional shape can be represented by the equation

$$\frac{x^2}{a^2} + \frac{y^2}{b^2} + \frac{z^2}{c^2} = 1.$$ (15.1)

We expect most elliptical galaxies to be either oblate ($a = b > c$) or triaxial ($a > b > c$). An oblate galaxy would appear circular, that is, would fall in class $E0$, if the line of sight is parallel to the z-axis. If viewed along the $x - y$ plane, it would appear elliptical, with semi-major and semi-minor axes equal to a and c, respectively. For other angles of observations, it would appear as an ellipse of different eccentricities.

These galaxies show practically no rotation and the stars move in random directions. Hence the flattened structure, observed in many elliptical galaxies, is not caused by rotation. This is in contrast to many astrophysical systems, such as the solar system, spiral galaxies, etc., whose disks are formed due to their rotation. The shape of the galaxy may be partially determined by the process of its formation and subsequent evolution, as discussed below. We can measure the distribution of velocities along the line of sight by observation of a spectral line profile. Let us consider a particular spectral line centered at frequency ν_0. The width of this line would be different for different stars. When we observe along the line of sight we obtain the integrated line profile for a large number of stars. The peak frequency for different stars is Doppler shifted due to their velocity component along the direction of observation. Hence the observed line profile is broadened due to the random motion of stars, as illustrated in Figure 15.6. The observed width of a spectral line, therefore, carries information about the distribution of velocities of different stars along the line of sight. This can be used to measure the velocity dispersion σ, that is, the standard deviation of the velocity component in the direction of observation.

One observes elliptical galaxies over a wide range of sizes and masses. The dwarf elliptical galaxies may be comparable to a globular cluster with a size of order 100 pc and contain about 10^7 stars. At the other end, we have ellipticals with size of order 100 Kpc and mass larger than 10^{13} solar masses. For example, the supergiant elliptical galaxy IC1101 is 2,000 times more massive compared to the Milky Way.

These galaxies are most likely formed by the merger of two or more spiral galaxies. Such collisions between different galaxies is much more likely in comparison to collisions between different stars because they are more tightly packed within galaxy clusters. When two galaxies collide, most of the stars simply pass through unobstructed due to their small size. Simulations show that as these galaxies merge, they tend to lose most of their gas. Furthermore, the pressure exerted by the two galaxies on one another triggers star formation. This tends to further disperse the gas and dust present in the merging galaxies. Hence at the end of this process, the final galaxy is relatively devoid of interstellar medium and further star formation is not possible. It is clear that the shape of the final galaxy will be determined by the nature of the merger between the two galaxies. Alternatively, elliptical galaxies may have formed due to the collapse of a cloud with very small angular momentum.

The surface brightness of ellipticals shows the following dependence on distance from the center,

$$I(r) = I_0 \exp\left(\alpha r^{1/4}\right), \qquad (15.2)$$

where $I(r)$ is the surface brightness at distance r from the center measured in units of L_{Sun}/pc^2, I_0 is the brightness at the center, and α is a constant. This is called de Voucouler's law. The parameter I_0 cannot be extracted very reliably due to constraints imposed by seeing. Hence it is more useful to express this law in a different form:

$$I(r) = I_e \exp\left(-7.67\left[\left(\frac{r}{r_e}\right)^{1/4} - 1\right]\right). \qquad (15.3)$$

Here r_e takes values in the range 1 to 10 Kpc. For a circular galaxy, r_e is the radius of the disk that emits half of the total radiation of the galaxy and I_e is the surface brightness at this distance. In general, I_e is the brightness at the isophote that encloses a region emitting half of the total luminosity of the galaxy.

The luminosity of elliptical galaxies shows an approximate dependence on their mean velocity dispersion σ given by,

$$L \propto \sigma^4. \qquad (15.4)$$

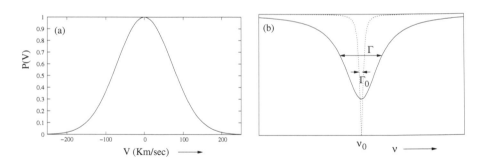

FIGURE 15.6: A spectral line appears broadened due to random velocities of stars in an elliptical galaxy. The left figure shows a typical velocity distribution, $P(V)$, of stars. Here V refers to the component of velocity along the line of sight. In Figure (b), the dashed line shows the absorption spectral line peaked at ν_0 for a particular star. The solid line shows the total spectral line due to all the stars along the line of sight. These stars have different velocity components along the line of sight and their peak positions are Doppler shifted, leading to a broadening of the line profile. The normalizations of both lines in (b) have been chosen arbitrarily.

This is called the Faber–Jackson relation. This relationship can be used to deduce the distance of these galaxies because their luminosities can be estimated by measurements of σ. This relationship follows from the virial theorem and the empirical fact that the surface brightness I_e at radius r_e of most elliptical galaxies is roughly the same. However, all elliptical galaxies do not obey this relationship.

15.2   Spiral Galaxies

The spiral galaxies contain a central bulge and a disk that extends over a much larger distance. The central bulge is similar in shape to an elliptical galaxy and has a very high concentration of stars. However, in contrast to an elliptical galaxy, the central bulge along with the entire galaxy undergo rotation. The disk thickness is typically about 1 Kpc and its diameter a few tens of Kpc. The spiral galaxies contain considerable amount of gas, which lies along a thin disk whose thickness is typically about 200 pc. The density of gas in this disk is nearly uniform. For example, the Milky Way is a spiral galaxy with a disk diameter of about 30 Kpc. The mass of gas and dust in the Milky Way is about 15% of its total mass in visible matter.

A striking feature of these galaxies is the spiral arms, which are located in their disks. These arms have a higher density of gas and dust and are sites of active star formation. They appear brighter in comparison to the rest of the disk due to the newly formed bright O, B stars. Hence spiral galaxies have

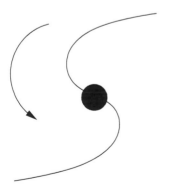

FIGURE 15.7: The sense of rotation of spiral galaxies in most cases is such that the spiral arms are trailing.

a much higher proportion of young stars in contrast to ellipticals, which are dominated by old stars. An image of a spiral galaxy is shown in Figure 15.3. The spiral structure suggests that the sense of rotation of these galaxies is most likely such that the spiral arms trail, as shown in Figure 15.7. However, a small percentage of these galaxies also show the opposite sense of rotation. The formation of spiral arms is best described by the density wave theory, discussed in Section 14.5.

Many spiral galaxies display a bar-like structure near the center. These are classified as barred spirals. An example of such a galaxy is shown in Figure 15.4. Many galaxies such as the Milky Way, which were earlier believed to be pure spirals, were later found to be barred spirals. Roughly two-thirds of all spirals are found to be barred. The spirals are found to be the most abundant type of galaxies. It is believed they slowly use up their gas and dust in the process of star formation and evolve into lenticular galaxies, S0 or SB0.

The surface brightness of the disk, $I(r)$, at a distance r from the center can be expressed as

$$I(r) = I_0 e^{-r/r_0} , \qquad (15.5)$$

where $r_0 =$ 1–5 Kpc and I_0 is the brightness at the center in units of luminosity per pc^2. Observationally the central brightness is given in units of magnitude per unit solid angle. For example, the B magnitude of the central region takes a wide range of values but is generally fainter than $B \sim 21.7$ mag/(square arc sec). As we discuss below, the total luminosity of these galaxies is related to their rotational speed.

Spiral galaxies show rapid rotation. The rotational speeds are deduced by measuring the Doppler shifts of the radiation received from different regions of the galaxy. The rotational speed of the Milky Way as a function of the distance, R, from the center is shown in Figure 15.8. The speed first increases and then becomes approximately constant as a function of R. The main features of this behavior are illustrated schematically in Figure 15.9. All spiral galaxies show this behavior. In the central region, the rotational speed increases approximately linearly with R. Hence this region rotates almost like a rigid body. However, at larger distances, the speed approaches a constant value. As discussed below, the variation in speeds with distance can be used to deduce the radial dependence of the mass of the galaxy.

We see in Figure 15.9 that the rotational speed V of spiral galaxies approaches a constant value at large distances. Let us denote this constant value by V_c. It turns out that the luminosity L of spiral galaxies is closely related to V_c. One finds

$$L \propto V_c^\alpha , \qquad (15.6)$$

where the exponent, α, takes a value close to 4. However, its precise value depends on the filter (B, V, etc.) used for flux measurements. This is called the Tully–Fisher relationship, after its discoverers. Clearly this relationship is very important because it allows astronomers to determine the luminosity of the distant galaxy by a measurement of its rotational speed. Hence the spiral galaxies act as standard candles, thus allowing determination of their distance.

Spiral galaxies are surrounded by a nearly spherically symmetric halo that

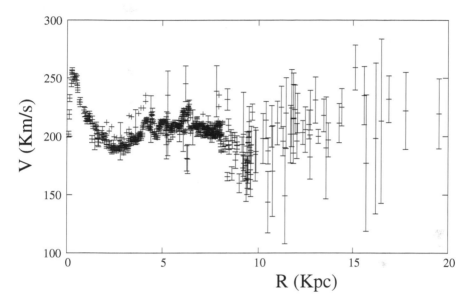

FIGURE 15.8: The observed rotational speed V of the Milky Way as a function of distance R from center. Data from "www.ioa.s.u-tokyo.ac.jp/~sofue/ h-rot.htm".

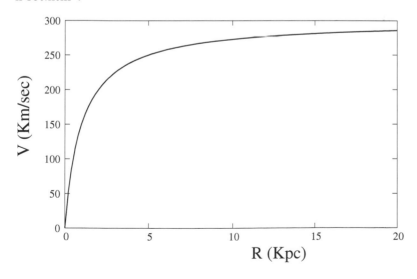

FIGURE 15.9: An illustration of the rotational speed V of spiral galaxies as a function of distance from the center.

extends much beyond the disk. Here the density of matter is very small in comparison to the disk. It is populated by globular clusters, which are huge clusters of very old stars, probably formed at the time of formation of the galaxy.

The region close to the center has a very high density of stars. It is believed that a very massive black hole resides at the center of all galaxies. The presence of a black hole must be deduced indirectly by inferring the total mass contained within some small radius. For example, in the case of the Milky Way, there is a very strong radio source, Sagittarius A*, at the center, which most likely contains a a supermassive black hole, with size many orders of magnitude larger than the mass of the Sun.

15.3   Evidence for Dark Matter

The dependence of the rotational speed of a spiral galaxy as a function of distance provides some information about its mass density. These observations provide evidence for dark matter in the Universe. By dark matter we mean matter that cannot be detected directly through observations of electromagnetic radiation at visible or other wavelengths. However, it makes its presence felt through its gravitational interactions. It is not observable directly because its interaction with radiation or normal matter, such as electrons and protons, is very weak. However, because it has mass and energy, it interacts gravitationally. We find that the observed rotational speeds of these galaxies cannot be explained in terms of the visible matter.

Let's consider the rotation of a star of mass m, located in the disk of a spiral galaxy, about the galactic center. We assume that it is located very far from the center, at a distance R. Let the rotational speed of the galaxy at this distance be V. The distance R is assumed to be large enough so that the density of visible matter beyond this point is almost negligible. In this case, the entire mass M of the galaxy is contained within this region $r < R$. For simplicity, we assume that R is large enough so that we can approximate M as a point mass. Applying Newton's law,

$$\frac{GMm}{R^2} = \frac{mV^2}{R}.$$
(15.7)

This implies that $V \propto 1/\sqrt{R}$, in disagreement with the observations that show that V is constant with R, as shown in Figure 15.9. The only way to explain this behavior is to postulate the existence of dark matter in the galaxy. We point out that a measurement of the rotational velocity of a spiral galaxy, at sufficiently large distance from the center, also gives an estimate of its mass M through Equation 15.7

Observations suggest that dark matter is not concentrated in the disk. It is generally assumed to have a spherically symmetric distribution. Let its density at distance r be denoted by $\rho(r)$. We next determine the $\rho(r)$ that leads to a flat rotation curve, that is, V equal to a constant, at large distances. In this analysis we ignore the contribution due to visible matter. For a test mass m at r, Newton's equation implies

$$\frac{GM(r)m}{r^2} = \frac{mV(r)^2}{r}, \tag{15.8}$$

where $M(r)$ is the mass of dark matter contained inside radius r. Because $V(r)$ is constant at large r, we find that

$$M(r) \propto r.$$

Differentiating this and using Equation 8.9, we find

$$\rho(r) \propto 1/r^2. \tag{15.9}$$

Hence we find that a spherically symmetric distribution of dark matter, whose density falls as $1/r^2$, is able to explain the observations.

15.4 Galaxy Clusters

Galaxies themselves are not seen in isolation but are found in groups or clusters. A group is a relatively small collection of galaxies. We are a part of a group of galaxies called the Local Group. It is a relatively small group whose diameter is equal to a few Mpc. It contains only three prominent members: the Milky Way, the Andromeda Galaxy (also called M31), and the galaxy M33. Besides this, the Local Group contains several smaller objects such as the Large Magellanic Cloud (LMC) and the Small Magellanic Cloud (SMC). Ignoring such small objects, Andromeda is the galaxy closest to the Milky Way. The Local Group is a part of the Local Supercluster, which is a supercluster of galaxies centered at the Virgo cluster of galaxies, located at a distance of 16.5 Mpc from us in the constellation Virgo. Most of the galaxies in the supercluster reside in a small number of clusters, which arrange themselves in a disk-like structure. A relatively smaller percentage of these clusters lie in a halo, which forms a roughly spherically symmetric structure. Most of the space in the supercluster is void. We observe a large number of clusters and superclusters of galaxies in the Universe. As we observe the Universe over distance scales much larger than 100 Mpc, we find that, on average, the number density of galaxies is the same at all positions and in all directions. Hence at such large distances, the Universe appears homogeneous and isotropic.

Exercises

15.1 Show that the total luminosity of a galaxy which satisfies the de Voucouler's law is given by

$$L \approx 7.2\pi I_e r_e^2 \,.$$

You may use

$$\int_0^\infty dx\, x^7 \exp(-x) = 7! \,.$$

15.2 We see only the two-dimensional projected view of the galaxies. Hence it is useful to define a two-dimensional density profile, after integrating over z coordinate, taken to be along the line of sight. Assume a spherically symmetric density profile, $\rho(r)$. Define the projected two-dimensional profile

$$\sigma(s) = \int_{-\infty}^\infty dz \rho(r) \,,$$

where $r^2 = s^2 + z^2$ and $s^2 = x^2 + y^2$. Determine $\sigma(s)$ for the density profile

$$\rho(r) = \frac{C}{r^2 + r_0^2} \,.$$

15.3 The density of visible matter in the Milky Way halo, at large r, can roughly be described by the formula

$$\rho(r) = \frac{C}{r^{3.5}} \,,$$

where C is a constant. What does this distribution predict for the r dependence of the rotation speed of the galaxy at large distances from the center? Show that this cannot explain the velocity profile at large distances from the center.

Chapter 16

Cosmology

Cosmology is based on the general theory of relativity (GTR). The GTR postulates that space-time is curved and dynamical. In Newtonian mechanics we think of space-time as a background stage on which particles reside, move and interact with one another. The space-time itself does not change with time. Furthermore, it is assumed that the three-dimensional space is described by Euclidean geometry. Such a space has no curvature. In contrast, Einstein's general theory of relativity postulates that space-time is curved. Furthermore, it is dynamical, that is, we cannot think of it as a background stage. The space-time itself evolves with time. The gravitational force is a manifestation of the space-time curvature. The evolution of space-time is determined by its matter content.

In order to understand what is meant by curvature, let us first take a simple example of a two-dimensional surface. A planar two-dimensional surface is flat, that is, has zero curvature. In contrast, the surface of a sphere is curved. One can understand the difference between the two by trying to roll a plain piece of paper into a spherical surface. It is simply not possible to do this. In contrast, one can easily roll that paper into a cylinder. Hence the surface of a cylinder is also flat.

We will introduce cosmology, assuming some of the basic results from the general theory of relativity. We will not discuss curved space-time and the general theory of relativity.

16.1 Euclidean Space

Let us first consider some properties of the Euclidean three-dimensional space. An important property is that we can use a Cartesian coordinate system throughout this space. This is the standard coordinate system, defined by three perpendicular axes, as shown in Figure 3.2, labeled as (x, y, z). Here we will instead label these coordinates as (x_1, x_2, x_3). Let us consider two points A and B in this space that have coordinates (x_1, x_2, x_3) and $(x_1 + dx_1, x_2 + dx_2, x_3 + dx_3)$, respectively. The coordinates of these two points are separated by an infinitesimal interval dx_1 along the x-axis, dx_2 along the y-axis, and dx_3 along the z-axis. Let us assume that these coordinate intervals are small, although for a flat space using Cartesian coordinates, this assumption is not needed. The distance ds between these two points is given by

$$ds^2 = dx_1^2 + dx_2^2 + dx_3^2. \tag{16.1}$$

This is also called the line element. A space in which we can use Cartesian coordinates, with the measure of distance given by Equation 16.1, at all points is called flat space.

Let us consider any two points in a flat space. We can always join these points by a straight line. This is the shortest path between the two points. The length of all other paths is longer. Another important property of a flat space is that the sum of the angles of a triangle is equal to 180°.

The Cartesian coordinates that we described above form a convenient coordinate system in flat space. However, one is free to use any other coordinate system that may be convenient. The new coordinates may be any arbitrary functions of the old coordinates. Hence we can define a new coordinate system (x_1', x_2', x_3') such that $x_i' = f_i(x_1, x_2, x_3)$ for $i = 1, 2, 3$. For different choice of coordinates, the coordinate separation between any two points will be different, that is, $dx_i \neq dx_i'$, in general. However the corresponding distance ds remains unchanged.

16.2 Curved Space

The simplest example of a curved space is the surface of a sphere. It is a two-dimensional space. We are interested in its intrinsic properties and not its embedding in the three-dimensional space. We can imagine a two-dimensional bug residing on this surface who knows nothing about the third dimension. We require a two-dimensional coordinate system which can be used to label all points on this surface. A two-dimensional Cartesian system (x_1, x_2) is simply not possible. One possible choice of coordinates is the standard latitudes and longitudes or equivalently the angular coordinates (θ, ϕ) where θ is the polar angle and ϕ the azimuthal angle.

The line element of a spherical surface can be written as

$$ds^2 = R^2(d\theta^2 + \sin^2\theta d\phi^2) \tag{16.2}$$

where R is a constant, identified as the radius of the sphere. It is not possible to make any transformation of the coordinates (θ, ϕ) that can convert this into the form, $ds^2 = dx_1^2 + dx_2^2$, over the entire surface. This is a general property of all surfaces or spaces which have non-zero curvature. We emphasize that the surface of a cylinder does not fall into this class. In order to determine whether a surface is curved one can compute a measure called the Gaussian curvature. We will not go into these details here. We only mention that for a spherical surface, the Gaussian curvature is equal to $1/R^2$. For a flat space the Gaussian curvature is zero.

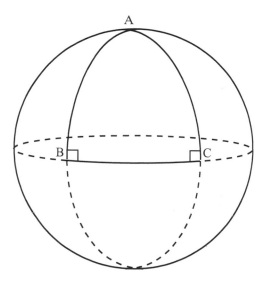

FIGURE 16.1: The triangle ABC is formed by three great circles on a spherical surface. These include the equator and two longitudes AB and AC. Clearly the longitudes meet the equator at right angles, as indicated at B and C. Hence the sum of the three angles of the triangle, $(\angle ABC + \angle BCA + \angle CAB) > 180°$. This indicates the curved nature of the surface. As explained in text, the great circles are the paths of shortest distance on a spherical surface. Hence these are equivalent to straight lines.

We next describe a few more properties of curved spaces. The shortest distance between any two points on a spherical surface is along the great circle joining these points. Hence if two points lie along a longitude then the shortest distance between them is the length of the (shorter) segment of the longitude joining them. If the points lie on different longitudes then one must first determine the great circle which joins them. The corresponding length of the segment of this circle gives the shortest distance between the points. Hence the great circles on a spherical surface are equivalent to straight lines on a plane. We may form a triangle on this surface by considering three points, A, B, C, which are joined together by great circles, as illustrated in Figure 16.1. The sum of angles of such a triangle is found to be greater than 180°, in contrast to a flat space where it is exactly 180°. This is another property of curved surfaces. The sum of the three angles of a triangle is not equal to 180°.

16.3   Minkowski Space-Time

Before we discuss cosmology, we review the Minkowski flat space-time. In the special theory of relativity we learn that space and time cannot be treated separately. A point in this space-time is called an event. Let us denote the time and space coordinates of an event by (t, x, y, z). The line element in this case is

$$ds^2 = -c^2 dt^2 + dx^2 + dy^2 + dz^2 \,, \tag{16.3}$$

where (dt, dx, dy, dz) represents the coordinate separation between two events infinitesimally close to one another and c is the speed of light, which is a universal constant in the special theory of relativity. It takes the same value in all frames of reference. As in the case of Euclidean space, ds is a measure of distance in this space. However, the reader may notice that ds^2 can take both positive and negative values, in contrast to the corresponding measure in Euclidean space. This property arises due to the negative sign of the time coordinate. We note that along the path of a light ray, $ds^2 = 0$. Such an interval is called light-like. The trajectory of any massive particle corresponds to $ds^2 < 0$. This interval is called time-like. The interval $ds^2 > 0$ is called space-like.

Consider a particle of rest mass m. In the special theory of relativity, the rest mass is defined as the mass of a particle when it is at rest. Let ds^2 be the interval between two points infinitesimally close to one another on the trajectory of a particle that may be at rest or in motion. Such an interval is necessarily negative. We define proper time τ such that $d\tau^2 = -ds^2 > 0$. Clearly, $d\tau$ is invariant under Lorentz transformations. It is equal to the time interval dt in the rest frame of the particle. This follows because $dx = dy = dz = 0$ in the rest frame. By the standard time dilation formula, we obtain $dt = d\tau/\gamma$ in a frame in which the particle is moving with speed v. Here, $\gamma = 1/\sqrt{1 - v^2/c^2}$, is called the boost factor.

16.4 Big Bang Cosmology

Modern cosmology is based on the hypothesis that the Universe is homogeneous and isotropic. This means that it appears the same at all points and in all directions. There is no preferred position or direction in the Universe. This hypothesis is called the cosmological principle. We note that the Universe in our vicinity certainly does not appear to satisfy this principle. For example, the solar system is centered at the Sun and forms a planar structure. As we go beyond the solar system, we observe galaxies and galaxy clusters, none of which appear isotropic or homogeneous. This principle is assumed to apply in a statistical sense on distance scales larger than 100 Mpc. For example, the mean density of matter, averaged over distance scales of order 100 Mpc, appears to be approximately the same in all directions and positions. This is the distance scale at which the Big Bang cosmology is applicable. We also point out that isotropy and homogeneity apply only in a particular frame called the cosmic frame of rest. An observer who is in motion with respect to this frame will not observe an isotropic distribution of matter. Due to Doppler and aberration effects, the density of matter would appear higher in the direction of motion, in contrast to other directions. A detailed explanation of this phenomenon is beyond the scope of this book. The Doppler effect also leads to an anisotropy in the temperature of the CMBR, as discussed later. Here we will always work in this preferred rest frame, unless stated otherwise.

16.4.1 Cosmological Redshift and Hubble's Law

The cosmological principle along with the general theory of relativity predicts that the Universe is expanding (or contracting) with time. This means that the physical distance between different galaxies is increasing with time. The expansion was first observed by Hubble in 1929. It was deduced by measuring the shift in wavelength of spectral lines from distant galaxies. One finds that the lines of distant galaxies have longer wavelengths. This change in the wavelength is expressed in terms of the parameter z, called the redshift. It is defined as

$$z = \frac{\lambda_o}{\lambda_e} - 1 \,, \tag{16.4}$$

where λ_e is the wavelength emitted by a distant galaxy and λ_o is the corresponding wavelength observed at Earth. Sources with redshifts as large as 6 have been observed. Because the galaxy is moving away from us, $\lambda_o > \lambda_e$ and $z > 0$. We can understand this in terms of the standard Doppler effect (Equation 3.9). If a source is moving away from the observer with speed v, then

$$\frac{\lambda_o}{\lambda_e} \approx 1 + \frac{v}{c} \,. \tag{16.5}$$

Hence we can interpret $z = v/c$. We clarify, however, that the Doppler effect generally refers to the shift in wavelength when the source is moving with respect to an observer in a flat space-time. In the present case, the effect is caused by the expansion of the space itself. Hence it is an effect caused by the gravitational force. The expansion is modeled in terms of the scale parameter $a(t)$, which increases as a function of time. Correspondingly, the physical distances between distant galaxies also increase.

The measurement of the redshift, z, can be made reliably by observing several different spectral lines. We can deduce a value of the redshift for each of these by comparing with the corresponding line observed in a laboratory transition. For consistency, all of these should give the same value of redshift, within experimental errors.

The observed redshifts, z, of distant galaxies show a direct dependence on the distance of the galaxy. This fundamental observation, called the Hubble's law, can be expressed as

$$z = \frac{H_0}{c} D, \qquad (16.6)$$

where D is the distance of the source and H_0 is called the Hubble constant. The observed value of this parameter is

$$H_0 = 67.80 \pm 0.77 \ (\text{km/s})/\text{Mpc}. \qquad (16.7)$$

The Hubble constant is historically written as

$$H_0 = 100h \ (\text{Km/s})/\text{Mpc}, \qquad (16.8)$$

where the parameter $h = 0.6780 \pm 0.0077$ absorbs the uncertainty in its value. The Hubble's law is often expressed in the form $v = H_0 D$, where $v \equiv cz$ is defined in Equation 16.5. We notice that a galaxy at a distance of 100 Mpc is expected to have a redshift of about 0.02.

The deduction of the Hubble's law crucially depends on the measurement of distances of galaxies. As we have seen in earlier chapters, a direct measurement is not possible beyond a distance of about 100 pc. For larger distances, one has to rely on standard candles, that is, sources whose luminosity can be deduced. As we discussed in Chapter 14, type Ia supernovae provide very good standard candles to deduce cosmological parameters.

Hubble's law shows that all galaxies are moving away from us. If we assume that the Universe is homogeneous, then observers located in other galaxies should also see all other galaxies moving away from them at the same rate as we observe. Hence we conclude that all galaxies are moving away from one another. Within the framework of the general theory of relativity, this arises due to the expansion of space. As time increases, the three-dimensional space expands, leading to an increase in the physical distance between the galaxies. This is illustrated in Figure 16.2. The dots in this figure are some representative galaxies. Their positions are measured with respect to a co-moving grid. The co-moving grid expands with time, such that the position of the galaxies relative to the grid remains fixed. Let the distance between two galaxies in such a grid be r. This distance is called the co-moving distance and remains fixed with time. The physical distance D between the galaxies is given by

$$D = a(t)r, \tag{16.9}$$

where $a(t)$ is the scale parameter. Hence D increases with time.

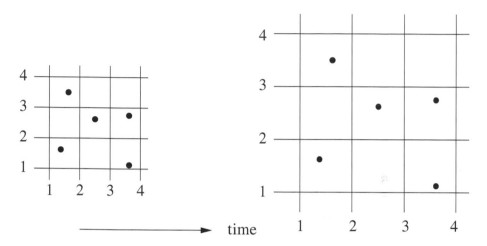

FIGURE 16.2: The physical distance between the galaxies, indicated by dots in this figure, increases proportionally to the scale factor $a(t)$ as the Universe expands. However the coordinate distance r remains fixed.

The galaxies may also have peculiar velocities. This refers to motion of galaxies besides the Hubble expansion. These arise due to local effects. The matter distribution in a small neighborhood of a galaxy is not homogeneous and isotropic. This leads to a net gravitational force that is responsible for peculiar motion. The peculiar velocities are relatively small on a cosmological scale and, to a good approximation, negligible. Due to their peculiar motion, galaxies move relative to the co-moving grid.

16.4.2 FRW Line Element

The Universe at very large distance scales is well described by the Friedmann-Robertson-Walker FRW line element, given by

$$ds^2 = -c^2 dt^2 + a^2(t)(dx_1^2 + dx_2^2 + dx_3^2) = -c^2 dt^2 + a^2(t)dr^2. \tag{16.10}$$

This gives the "distance" ds between two events that have coordinate separation equal to (dt, dx_1, dx_2, dx_3). This differs from the Minkowski space-time by the presence of the scale parameter $a(t)$, discussed above. The curvature of such a space-time is non-zero. However, in this particular case, the three-dimensional space, at any fixed time, has zero curvature. Cosmological observations suggest that the Universe is flat, that is, the space part has zero curvature. The coordinates x_1, x_2, and x_3 are called co-moving coordinates. The co-moving coordinates of any object do not change with time, as discussed above. As in the case of the Minkowski line element, $ds^2 = 0$ along the trajectory of a light ray. Hence in a small time interval dt, a light ray travels a co-moving distance dr given by

$$dr = c\,dt/a(t).$$ (16.11)

We point out that the physical distance D, given in Equation 16.9, is the distance two galaxies at some fixed time. It is given by, $D = \int ds$, where ds is computed at any time by setting $dt = 0$ in Equation 16.10.

FIGURE 16.3: The wavelength of radiation increases in proportion to the scale parameter as the Universe expands.

We next relate the scale parameter to the observed redshift of a distant galaxy. Let us assume that a distant galaxy emits radiation of wavelength λ_e at time t_e, as shown in Figure 16.3. The rate of expansion is assumed to be very small so that the scale parameter does not change significantly over one time period of the wave. At a later time t_o, this radiation is observed by the observer O. Due to the expansion of the Universe, its wavelength at this time is λ_o. Let the time interval between the emission of two wave crests by the galaxy be equal to dt_e and the time interval between these crests according to the observer be dt_o. By Equation 16.10, $dt_o/dt_e = a(t_o)/a(t_e)$. Here we have used the fact that the co-moving distance between two wave crests does not change. The frequency of light is inversely proportional to the time period. Hence the frequency of the emitted wave, ν_e, is related to that of the observed wave, ν_o, by the formula

$$\frac{\nu_e}{\nu_o} = \frac{a(t_o)}{a(t_e)}.$$ (16.12)

The speed of light remains the same at all times. Hence we obtain

$$\frac{\lambda_o}{\lambda_e} = \frac{a(t_o)}{a(t_e)}.$$ (16.13)

We should have expected this result because all physical distances expand in proportion to the scale parameter. We will take t_o as the current time and set $a(t_o) = 1$. Using Equation 16.4 we obtain

$$1 + z = \frac{1}{a(t)}, \tag{16.14}$$

where we have set $t_e = t$.

We next establish Hubble's law using the Doppler shift formula, Equation 16.5, valid for small v/c. The speed of a distant galaxy can be obtained by taking the time derivative of Equation 16.9. We obtain

$$v = \dot{a}(t)r = \frac{\dot{a}}{a}D$$

which is Hubble's law with $v = cz$ and the Hubble parameter

$$H(t) = \frac{\dot{a}}{a}. \tag{16.15}$$

The Hubble parameter $H(t)$ is, in general, a function of time. Its value at current time is denoted as H_0 and is given by Equation 16.7.

16.4.3 Matter and Radiation

The rate of expansion is determined by the energy density and nature of the matter present. The common type of matter we are familiar with is formed of atoms and molecules, which typically have speeds much smaller than the speed of light. Such matter is called nonrelativistic. Its energy density is dominated by its rest mass energy. A particle of rest mass m has energy equal to mc^2. Hence if the mass density of matter is ρ_m, then its energy density ρ_e is equal to $\rho_m c^2$. This applies to all forms of familiar matter, which may be solid, liquid, or gas. We are interested in matter density averaged over very large distance scales, on the order of 100 Mpc. The number density of galaxies is approximately $0.1/(\text{Mpc}^3)$. Hence a volume of $(100 \text{ Mpc})^3$ contains on the order of 10^5 galaxies. For the purpose of determining the overall expansion of the Universe, we will assume uniform density and ignore fluctuations around its mean value.

For cosmological purposes, radiation is also considered a form of matter. The term *radiation* includes photons as well as other particles that may have speeds close to the speed of light. Such particles are called relativistic. At present, the total energy density contributed by radiation is very small. However, at sufficiently early time, the temperature of the medium was very high and radiation dominated the energy density of the Universe.

We will assume that matter is approximately at equilibrium and assume the perfect gas equation of state. The equation of state of a perfect gas of radiation or relativistic particles is given by Equation 8.53. We express it as

$$P_R = \frac{\rho_R}{3} \quad \text{(radiation or relativistic matter)}, \tag{16.16}$$

where P_R and ρ_R, respectively, denote the pressure and energy density of radiation. As the Universe evolved, it cooled. Eventually the nonrelativistic matter started to dominate. In this case the pressure is negligible compared to the energy density. In this limit, energy density gets a dominant contribution from mass energy. Let ρ_{NR} and P_{NR} denote, respectively, the energy density and pressure of nonrelativistic matter. We obtain $\rho_{NR} \approx nmc^2$, where n is a number density of particles and m their mass. The familiar nonrelativistic matter is formed of atoms or molecules, which themselves are formed of electrons, protons and neutrons. The dominant contribution to ρ_{NR} comes from protons, and neutrons, whose mass is much larger than that of electrons. Hence $m \approx m_p \approx m_n$, where m_p and m_n denote the masses of the proton and neutron, respectively. The pressure is given by the ideal gas law. We obtain $P_{NR} = nm < v^2 > /3$ (Equation 8.41), where kT has been replaced in terms of $< v^2 >$ using Equation 8.69. It is clear that $P_{NR} << \rho_{NR}$. Hence we can approximate it as

$$P_{NR} \approx 0 \quad \text{(nonrelativistic limit)}. \tag{16.17}$$

We next determine how the energy density of these different components changes as the Universe expands. For nonrelativistic matter, this is very simple. The amount of matter does not change. The volume of space increases as $a(t)^3$. Hence both the mass density and the energy density decrease as $1/a^3$. We can express this as

$$\rho_{NR} = \frac{\rho_{NR0}}{a^3}, \tag{16.18}$$

where ρ_{NR0} denotes the current value of the nonrelativistic energy density. For the case of radiation, the wavelength of photons increases in proportion to $a(t)$. The energy of a photon is inversely proportional to its wavelength and, hence, the energy per photon decreases as $1/a(t)$. Furthermore, the number density of photons decreases as $1/a^3$ due to an increase in the volume of the Universe. Hence the energy density of radiation decreases as $1/a^4$, that is,

$$\rho_R = \frac{\rho_{R0}}{a^4}, \tag{16.19}$$

where ρ_{R0} is its current value.

Before proceeding further, we point out that the equations given above already reveal some fascinating facts about the Universe. As $a(t)$ increases, the energy density of nonrelativistic matter decreases as $\rho_{NR} \propto 1/a^3$ and that of relativistic matter decreases at a faster rate, $\rho_R \propto 1/a^4$. At the current time, ρ_R is very small. However, because it decreases faster with time, it follows that at sufficiently early times, it must dominate. The energy density of radiation is proportional to T^4 (see Equation 4.21). Hence, at an early time, when the Universe was dominated by radiation, its temperature decreased as

$$T \propto 1/a(t). \tag{16.20}$$

Hence we deduce that the Universe was hotter at early times. Furthermore, even nonrelativistic components, which were in thermal equilibrium with radiation at early times, had the same temperature and cooled at the same rate as the Universe expanded. At a certain stage, matter (i.e., electrons and protons) and radiation went out of equilibrium. A little beyond that time, atoms formed and eventually led to the formation of structures in the Universe.

We can now explain the origin of the name Big Bang. Because the Universe is expanding, it must have been very small at early times. It was also very dense and hot. Extrapolated sufficiently back in time, the size becomes almost zero and the density and temperature extremely large. Hence we can imagine that the Universe originated in an event called the Big Bang, which attributed it with such high temperature and density. We can add that there is no real understanding of the actual nature of this phenomenon. We are only able to reliably describe the sequence of events starting a little after the Big Bang.

16.4.4 Cosmological Evolution Equations

Cosmological observations suggest that, to a good approximation, the Universe is flat, that is, the three-dimensional space has zero curvature. In fact, this approximation holds for almost the entire evolution of the Universe. The space is flat, that is, has zero curvature, if the total energy density is equal to the critical density,

$$\rho = \rho_{cr} . \tag{16.21}$$

If $\rho > \rho_{cr}$, then the curvature is positive and $\rho < \rho_{cr}$ leads to negative curvature. We will use the symbol ρ_{cr} for the value of critical energy density at current time. It is convenient to define the parameter Ω such that

$$\Omega = \rho/\rho_{cr} , \tag{16.22}$$

that is, it is the ratio of the energy density of the Universe at any time to the critical energy density today. The value of this parameter at current time is denoted by Ω_0. Since the energy density at present is very close to ρ_{cr}, we expect that

$$\Omega_0 \approx 1 . \tag{16.23}$$

Assuming a flat Universe, the differential equations for the scale parameter can be written as

$$\left(\frac{\dot{a}}{a}\right)^2 = \frac{8\pi G}{3c^2}\rho \tag{16.24}$$

$$2\frac{\ddot{a}}{a} + \left(\frac{\dot{a}}{a}\right)^2 = -\frac{8\pi G}{c^2}P , \tag{16.25}$$

where P is the pressure. These equations can be derived using Einstein's equations of general relativity, which we shall skip. In Exercise 16.1, we develop a model of expansion based on Newtonian mechanics. We next use these equations in order to derive some properties of the expanding Universe. These equations are valid if the energy density is equal to the critical value. Hence, using Equation 16.24 we can obtain the value of the critical density today, ρ_{cr}. We find

$$\rho_{cr} = \frac{2H_0^2 c^2}{8\pi G}, \tag{16.26}$$

where we have used Equation 16.15. Using the value of the Hubble constant H_0, we obtain the critical mass density,

$$\frac{\rho_{cr}}{c^2} = 1.87h^2 \times 10^{-29} \text{ g/cm}^3. \tag{16.27}$$

This is the current density of matter favored by cosmological observations. However, the density of visible matter in the Universe is far lower than this. This again indicates the presence of dark matter in the Universe. Recall that dark matter is also needed in order to explain the rotation curves of spiral galaxies.

We next solve Equations 16.24 and 16.25 in order to determine the time evolution of $a(t)$. For nonrelativistic matter, using Equations 16.17 and 16.18, we obtain

$$a(t) = \left(\frac{t}{t_0}\right)^{2/3}, \tag{16.28}$$

where t_0 is the present time of the Universe and we have set $a(t_0) = 1$. For relativistic matter, we obtain

$$a(t) \propto (t)^{1/2}. \tag{16.29}$$

We notice that both for the case of nonrelativistic matter and radiation dominated Universe, $\ddot{a} < 0$. Hence, the rate of expansion decelerates in both of these cases.

16.4.5 Accelerating Universe and Dark Energy

Let us assume that for most of its lifetime, the Universe was dominated by nonrelativistic matter, including dark matter. The time interval over which radiation dominated at early times is very small. Hence in Equation 16.28, t_0 can be interpreted as the lifetime of the Universe. By differentiating Equation 16.28 with time, we can relate t to the Hubble parameter. This leads to an estimate of t_0 given by

$$t_0 = \frac{2}{3H_0}. \tag{16.30}$$

Using the value of H_0, this gives $t_0 = 9.6 \times 10^9$ years or 9.6 billion years. This value turns out to be too small. The age of the globular clusters in the Milky Way is estimated to be about 12.7 billion years. Hence something is wrong with our model. The globular clusters cannot be older than the Universe. It turns out that this model, in which we assume dominance by nonrelativistic matter, also conflicts with other cosmological observations.

The cosmological parameters can be probed somewhat directly using data from supernovae type Ia. They have now been observed over very large distances. The farthest supernova observed so far corresponds to a redshift equal to 1.7. At such large redshifts, the simple relationship given in Equation 16.6 is no longer valid. In order to get a suitable relationship between distance and redshift, one has to first define an observable measure of the distance. Recall that if L is the luminosity of a source, then its flux F at a distance r is given by

$$F = \frac{L}{4\pi r^2} \, . \tag{16.31}$$

This equation is valid in flat space-time. In the present case, the relationship between flux and luminosity is somewhat more complicated because the space-time is curved. Let us define a distance, called the luminosity distance d_L, such that

$$F = \frac{L}{4\pi d_L^2} \, . \tag{16.32}$$

All complications due to the expansion of space are dumped into d_L. The astronomers can measure the flux of a distant source. Hence we can get an estimate of d_L if we can deduce its luminosity.

We next relate d_L to the redshift of a distant galaxy. Consider a galaxy located at a co-moving distance r. Assume that it emits a light wave at time t_e that is observed at time t_o. As discussed earlier, the co-moving distance, dr, traveled by a light wave in a small time interval dt is equal to $cdt/a(t)$. Hence, the distance r can be expressed as

$$r = \int_{t_e}^{t_o} \frac{cdt}{a(t)} \, . \tag{16.33}$$

Let us assume that the galaxy emits photons isotropically in time interval dt_e. On a co-moving grid, these photons are distributed over an area $4\pi r^2$ at a co-moving distance r. Hence the flux observed is given by

$$F \propto \frac{L}{4\pi r^2} \, . \tag{16.34}$$

Furthermore, each photon is redshifted. Hence its energy is reduced by a factor $a(t_e)/a(t_o)$. The photons emitted in time interval dt_e will spread out due to expansion and will be observed in a time interval dt_o, such that $dt_o/dt_e = a(t_o)/a(t_e)$. Hence the flux, that is, the energy observed per unit time per unit area, is equal to

$$F = \frac{La^2(t_e)}{4\pi r^2} \, , \tag{16.35}$$

where we have set $a(t_o) = 1$. Comparing with Equation 16.32, we obtain the luminosity distance

$$d_L = \frac{r}{a(t_e)} = (1+z)r \, . \tag{16.36}$$

We next express the formula for r in terms of an integral over the redshift. We first change variables in Equation 16.33 from t to a using

$$da = aH dt \,,$$

where H is the Hubble parameter. Using Equation 16.14, we change variables from a to z. We obtain

$$r = \int_0^z \frac{c\,dz'}{H(z')} \,. \tag{16.37}$$

In order to proceed further, we need $H(z)$, which depends on the nature of the matter filling up the Universe. Let us assume that the energy density of the Universe is dominated by nonrelativistic matter. We rewrite Equation 16.24 as

$$H^2 = \frac{8\pi G}{3c^2} \rho \,.$$

Dividing by H_0^2, using Equation 16.26, we obtain

$$H(t) = H_0 \sqrt{\Omega} \,, \tag{16.38}$$

where Ω is given by Equation 16.22. For the case of the nonrelativistic matter dominated Universe, $\rho = \rho_0/a^3 = \rho_0(1+z)^3$. Hence we set

$$\Omega = \Omega_M = \Omega_{M0}(1+z)^3 \,,$$

where

$$\Omega_{M0} = \frac{\rho_0}{\rho_{cr}}$$

is the value of Ω_M today. For a flat, nonrelativistic matter dominated Universe, $\Omega_{M0} = 1$. Hence we obtain

$$r = \int_0^z \frac{c\,dz'}{H_0(1+z)^{3/2}} = \frac{2c}{H_0} \left[1 - \frac{1}{(1+z)^{1/2}} \right] \,. \tag{16.39}$$

Using Equation 16.36 we obtain the formula for d_L for this model. We can determine the value of H_0 by fitting to the observed flux for supernovae type Ia.

In Figure 14.3, we show the light curve of a typical type Ia supernova at $z = 0$. It shows the variation of intensity in the V band as a function of time. As discussed in Chapter 14, the peak luminosity of type 1a supernovae is approximately the same for all supernovae. One does find some spread in the peak luminosity. It turns out that the peak luminosity is correlated with the width of the light curve. A supernova with larger width has larger peak luminosity. This relationship can be quantified and then used to obtain a reliable estimate of the luminosity of a distant supernova. The extracted luminosity allows us to extract the absolute magnitude M corresponding to some filter. The apparent magnitude m corresponding to this filter can be extracted by flux measurement. The difference, $m - M$, is the distance modulus. One typically measures the distance modulus corresponding to the standard astronomical filters. Note that the light from a distant supernova is redshifted. Hence if the flux of a nearby supernova peaks at blue frequency, a similar supernova at large distance may show a peak closer to red frequency.

The fit of the cosmological model based on nonrelativistic matter to the supernova data, shown in Figure 16.4, turns out to be not very good. The goodness of the fit can be quantified by computing a statistical quantity called χ^2 (chi-square). We will not go into these details here. The important point is that the distant supernovae appear to be dimmer, that is, have a larger distance modulus, in comparison with what is implied by this fit. Hence these supernovae are probably further away. These observations were rather unexpected and presented a major challenge to the Big Bang cosmology. It was found that a very good fit is obtained if one introduces an additional parameter, called the cosmological constant, into Einstein's equations. However, most cosmologists believe that a proper understanding of this phenomenon is still lacking.

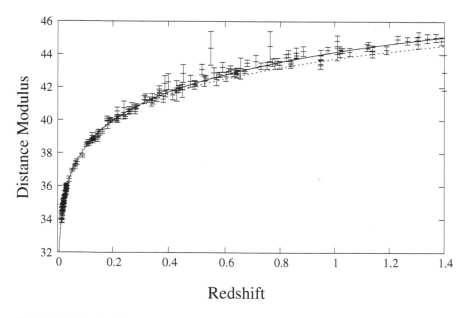

FIGURE 16.4: The fit to supernovae type Ia using a nonrelativistic matter dominated (dotted line) model and a model that includes both nonrelativistic matter and a cosmological constant (solid line). (Data from Amanullah et al., *Astrophysical Journal* 716, 712 (2010).)

Including the cosmological constant, denoted by Λ, the evolution equations for $a(t)$ become

$$\left(\frac{\dot{a}}{a}\right)^2 = \frac{8\pi G}{3c^2}(\rho + \Lambda), \tag{16.40}$$

$$2\frac{\ddot{a}}{a} + \left(\frac{\dot{a}}{a}\right)^2 = -\frac{8\pi G}{c^2}(P - \Lambda). \tag{16.41}$$

Notice that the cosmological constant can be treated as a new form of matter whose energy density and pressure are related by the formula

$$P = -\rho, \tag{16.42}$$

with $\rho = \Lambda$. This implies that the cosmological constant has negative pressure. Physically one can associate the cosmological constant with vacuum energy. Recall that in mechanics, the absolute value of potential energy is arbitrary. We usually arbitrarily fix the potential energy at some point. Once this choice is made, the potential energy for any physical system is fixed uniquely. This also implies that the minimum value of the potential energy is arbitrary. This value is exactly what we call vacuum energy. It is the energy (not including the rest mass energy) that a system of particles would have if all the particles had zero kinetic energy and were located at the minimum of the potential energy. In quantum mechanics, it is the energy of the system in its ground state. This vacuum energy plays no role in physics as long as we ignore gravity. However, all forms of energy including the vacuum energy contribute to the gravitational force, or equivalently, the curvature of space-time. Hence the vacuum energy also contributes to cosmological expansion.

It is thus far not clear whether the additional component required to fit the cosmological observations is indeed the cosmological constant. However, it is clear that it must have negative pressure. Hence cosmologists assume the existence of a new component, which is called dark energy, whose equation of state is given by

$$P = w\rho, \tag{16.43}$$

with $w < 0$. The cosmological constant or vacuum energy is a particular example of this with $w = -1$. For the case of pure cosmological constant, with negligible radiation or matter density, the scale factor is given by

$$a(t) = a_0 \exp\left[\sqrt{\frac{8\pi G\Lambda}{3c^2}}\, t\right], \tag{16.44}$$

where a_0 is a constant. The term *dark* suggests that this component has so far not been observed directly. Furthermore, we know even less about its properties in comparison to those of dark matter. The main property of dark energy is that if $w < -1/3$, it leads to accelerating Universe ($\ddot{a} > 0$). This is easily verified by substituting Equation 16.43 into Equation 16.25. This is in contrast to what we found for familiar nonrelativistic matter or radiation.

We next describe the cosmic evolution assuming the presence of both dark energy and nonrelativistic matter. The nonrelativistic matter contains both dark matter and visible matter. For dark energy we will set $w = -1$. In this case, Equation 16.38 modifies to

$$H(t) = H_0\sqrt{\Omega_\Lambda + \Omega_M} = H_0\sqrt{\Omega_\Lambda + \Omega_{M0}(1 + z)^3}, \tag{16.45}$$

where Ω_Λ is the additional contribution due to the cosmological constant. It is simply a constant, independent of redshift. Using Equation 16.23 we obtain

$$\Omega_\Lambda + \Omega_{M0} = 1 \,.$$

The fit to supernova data based on Equation 16.45 is also shown in Figure 16.4. We find that this gives a very good fit, supporting the presence of a cosmological constant or dark energy. The best fit parameters are $H_0 = 69.9$ (Km/s)/Mpc, $\Omega_M = 0.27$, $\Omega_\Lambda = 1 - \Omega_M$. Hence we conclude that the energy density of the Universe is composed of 73% cosmological constant and 27% nonrelativistic matter. The nonrelativistic matter is predominantly dark matter. The visible matter contributes less than 5% of the total energy density. This model of the Universe, based on a cosmological constant, is called the Lambda Cold Dark Matter (LCDM) model. The value of H_0 obtained by fitting the supernova data is close to the value given in Equation 16.7. The difference arises because the best value, Equation 16.7, also uses other cosmological data such as the CMBR.

Remarkably, the LCDM model also solves the problem of the age of the Universe. Using Equation 16.45, we obtain the following formula for the age of the Universe:

$$t_0 = \frac{1}{H_0} \int_0^1 \frac{da}{a} \frac{1}{\sqrt{\Omega_\Lambda + \Omega_M/a^3}} \,. \tag{16.46}$$

By performing the integral numerically, we obtain $t_0 = 14.3$ years where we have used Equation 16.7 for the Hubble constant. This value of the age is consistent with the age of the globular clusters.

16.5   The Early Universe

At early times, the Universe was hot, with its energy density dominated by radiation. At the time corresponding to the Big Bang, the temperature was perhaps such that $kT \sim M_P c^2$, where $M_P \sim 10^{19}$ GeV/c^2 is the Planck mass and k the Boltzmann constant. The physics at such a high energy scale is not well known and we are not able to describe the evolution of the Universe at this time very well. Soon after, it went through a phase of very rapid expansion, called inflation. Due to rapid expansion the Universe became very cold. However, after the end of inflation it went through a phase in which it was reheated and again became dominated by radiation. The temperature at this stage was very high, corresponding to an energy scale of order $kT \sim 10^{15}$ GeV. In order to get a handle on this scale, recall that $kT \sim 1$ eV corresponds to a temperature of approximately 10,000K. Hence the temperature after reheating was very, very high. As this high temperature, the Universe was composed of plasma that may have contained known particles such as electrons, photons, along with many exotic particles that have so far not been seen experimentally. During this phase the evolution is described by Equation 16.29. The physics of the Universe at $T \sim 10^{15}$K, corresponding to an energy scale $kT \sim 100$ GeV, is well described by the Standard Model of particle physics. At this temperature, the plasma would be composed of charged leptons (e^-, μ^-, τ^-), neutrinos $(\nu_e, \nu_\mu, \nu_\tau)$, quarks (u,d,s,c,b), photons, gluons, $W^\pm$, Z, and their antiparticles. The top (t) quarks would have decayed by this time.

At $T \sim 10^{15}$K, quarks and gluons existed as free particles. As the Universe cooled further to $T \sim 10^{12}$K, a phase transition occurred and free quarks and gluons ceased to exist. They became confined inside hadrons, that is, particles such as protons and neutrons. At this stage, the plasma was composed of photons, electrons, muons, neutrinos, protons, and neutrons. The remaining particles would have either decayed or formed bound states. The antiparticles of electrons, muons, and neutrinos would also be present. The antiparticles of protons (and neutrons) would have disappeared due to annihilations with the corresponding particles to form radiation, by the process

$$p^- + p^+ \rightarrow \gamma + \gamma. \tag{16.47}$$

The temperature at this time is such that $kT << m_b c^2$, where $m_b \approx m_p \approx m_n$ denotes the mass of proton, m_p, or neutron, m_n. The subscript 'b' stands for baryon. Hence photons do not have enough energy to create proton-antiproton pairs by the inverse process corresponding to Equation 16.47. In contrast, photons do have enough energy to create electron-positron pairs and, hence, the process

$$\gamma + \gamma \leftrightarrow e^- + e^+ \tag{16.48}$$

proceeds in both directions. This maintains both electrons and positrons in thermal equilibrium. The inverse of this process is called the pair annihilation process. In equilibrium, both of these processes contribute at the same rate. We point out that protons and neutrons are nonrelativistic at this temperature, whereas electrons, muons, and neutrinos are relativistic.

We next describe a very interesting prediction of the Big Bang cosmology. The process of proton-antiproton annihilation, Equation 16.47, dumps a huge number of photons in the Universe. This is because at very early time the density of protons and antiprotons was nearly equal, with a small excess of protons. This small excess represents a small asymmetry between particles and antiparticles. Hence after the annihilation process, only a small number of protons remained. All the remaining proton-antiproton pairs converted to photons. We point out that if the number of particles and antiparticles were equal, there would be no matter left today. The presence of matter indicates a small excess of particles over antiparticles. Hence the pair annihilation process predicts a very large number density of photons today in comparison to protons. As we will discuss, this is in good agreement with the data.

The mass of a muon, M_μ, is such that $M_\mu c^2 \sim kT$, with $T \sim 10^{12}$K. As the temperature falls significantly below this value, muon-antimuon pairs annihilate into photons. The pair creation process can no longer contribute. At the end of this annihilation process, a small number of muons remain, as in the case of protons. Again, this is because of a small excess of muons over antimuons present since very early times. As the Universe evolved further in time, at some stage these muons decayed to electrons and neutrinos, and disappeared from the cosmic plasma.

Let us now consider the time when the temperature was slightly higher than 10^{10}, corresponding to $kT \sim 1$ MeV. At this temperature the scattering rate, Γ, of neutrinos with the plasma falls below the expansion rate, that is, $\Gamma < H(t)$. Beyond this point, neutrinos fall out of equilibrium with the plasma and evolve independently. Due to the small scattering rate, their mean free path becomes very large, comparable with the size of the Universe. This follows from the inequality $\Gamma < H(t)$. Notice that $\tau = 1/\Gamma$ is the time interval between two collisions and τc the mean free path. Furthermore, $1/H$ is the typical time scale of expansion of the Universe. It is the time over which its size changes significantly. Hence c/H provides a rough estimate of the observable size of the Universe. The constraint $\Gamma < H(t)$ implies $c\tau > c/H$, which is the result claimed above. Hence, beyond this time, neutrinos rarely interact with the plasma particles. Similarly, all other particle species decouple from the cosmic plasma when their reaction rate $\Gamma < H(t)$.

At the time of decoupling the neutrinos were relativistic particles described by the Fermi-Dirac distribution. After decoupling their distribution remains the same with $T \propto 1/a(t)$ (see Equation 16.20). The photons remain coupled to the plasma at this time and decouple much later. Their temperature also decreases at the same rate. The photon gas is observed today as the cosmic microwave background radiation (CMBR) with a blackbody temperature of about 2.73K. Because both neutrino and photon temperatures evolve similarly, we expect the neutrino gas temperature to be also equal to 2.73K. This is roughly correct with one small correction. After the neutrinos decoupled, the electron-positron pairs annihilated, dumping their entire energy into the cosmic plasma. This led to a significant enhancement of the plasma temperature. The neutrinos that had already decoupled did not get any contribution from this. Hence their temperature evolved strictly as $1/a$, whereas the photon gas got heated during this process. This leads to a small difference in temperature between neutrino and photon gases today.

A detailed calculation predicts that the neutrino temperature T_ν today is related to the corresponding photon temperature T_γ by the formula

$$\frac{T_\nu}{T_\gamma} = \left(\frac{4}{11}\right)^{1/3} . \tag{16.49}$$

We will skip the proof of this relationship. With $T_\gamma = 2.73$K, we find that $T_\nu = 1.94$K. It has not been possible so far to observe the primordial neutrino gas. This is because neutrinos interact very weakly with matter and are very difficult to observe. This remains one of the observational challenges of modern cosmology.

As mentioned above, the electron and positron pairs annihilated after neutrino decoupling, producing photons through the pair annihilation process,

$$e^- + e^+ \rightarrow \gamma + \gamma \,. \tag{16.50}$$

In analogy with proton-antiproton annihilation, this phenomenon happens once the energy scale $kT \ll m_e c^2$, where m_e is the electron mass. At such low energies, the inverse process, Equation 16.48, is no longer effective. At the end of this process, a small number density of excess electrons remains.

Furthermore, the fact that at early times, matter and antimatter were present in almost equal quantities explains the very large number density of photons in comparison to protons observed today. At various stages during the evolution of the early Universe, different pairs of particles and antiparticles annihilated to create radiation. This process left behind a small fraction of particles, which we see today as protons, electrons, and neutrons, either in the free or bound state. The number of particles present today is equal to the difference between the number of particles and antiparticles at early times. We expect this difference to be very small because the theory of fundamental particles is almost symmetric with respect to the interchange of particles with antiparticles. This is called the charge conjugation symmetry. This also predicts that the number density of photons, produced by the annihilation process, should be very large at the current time. Let n_e and $\bar{n}_e$ represent the number densities of electrons and positrons, respectively, at early times. We expect

$$\left. \frac{n_e - \bar{n}_e}{n_e} \right|_{early} \ll 1 \,. \tag{16.51}$$

At early time, when electrons and positrons were relativistic, n_e is approximately equal to n_γ. Hence it is clear that, at the end of the annihilation process, $n_e/n_\gamma \ll 1$ because all the pairs would annihilate to produce photons. Let us assume that the number densities, n_e and n_γ decreased with further expansion of the Universe at the same rate. In that case the present observed ratio of matter particles to photons should be of the same order as the ratio on the left hand side of Equation 16.51.

Let n_b represent the current number density of baryons, which includes protons and neutrons. Notice that the number of protons and electrons today must be almost equal to one another because we expect that the Universe is electrically neutral. Furthermore, the number of protons is of the same order as the number of neutrons in the Universe. Hence n_b is roughly equal to the current number density of electrons. This implies that the ratio

$$\eta = n_b/n_\gamma \,, \tag{16.52}$$

observed today would be approximately the same as the ratio in Equation 16.51. Let us now estimate this ratio. The number density of baryons in the Universe is simply equal to the total energy density of visible matter divided by the mass of the proton. This is because the mass of electrons is negligible. The energy density of visible matter is roughly 2% of the critical energy density. Hence we expect that

$$\Omega_b = \frac{\rho_b}{\rho_{cr}} \approx 0.02 \,, \tag{16.53}$$

where ρ_b denotes the energy density of baryons. This implies

$$n_b \approx \frac{\rho_b}{m_p c^2} = \frac{\rho_{cr}}{m_p c^2} \Omega_b \,. \tag{16.54}$$

The photon number density is obtained using the Bose–Einstein distribution. We have

$$n_\gamma = \frac{g_{int}}{2\pi^2 \hbar^3} \int_0^\infty \frac{p^2 dp}{\exp(E/kT) - 1} \,, \tag{16.55}$$

where for photons, $g_{int} = 2$ and the chemical potential $\mu = 0$. We obtain

$$n_\gamma = \frac{2}{\pi^2} \left(\frac{kT}{c\hbar} \right)^3 \zeta(3) \,, \tag{16.56}$$

where $\zeta(3) \approx 1.202$ is the Riemann zeta function. Using Equations 16.56, 16.54, and 16.26, we obtain

$$\eta = 5.4 \times 10^{-10} \frac{\Omega_b h^2}{0.02} \,. \tag{16.57}$$

Hence, as expected, we find that the number density of photons is very large in comparison to that of protons or electrons. As we discuss below, this has important implications for several processes in the early Universe, such as primordial nucleosynthesis, formation of atoms, etc.

16.5.1 Primordial Nucleosynthesis

One of the great successes of the Big Bang model is that it correctly explains the abundance of light elements in the Universe. As we learned in Chapter 9, elements up to iron are synthesized in stellar interiors. Elements heavier than iron are also synthesized in stars under special conditions. However, stellar nucleosynthesis is not able to properly explain the abundance of light elements, such as ^{4}He, ^{3}He, D, ^{7}Li, in the Universe. Let us first consider ^{4}He. As we learned in Chapter 9, helium is produced in stellar interiors during the main sequence phase. However, most of it either is converted to heavier elements during the giant phase or stays confined inside white dwarf stars. It is not released into the interstellar medium. Hence if the density of helium in this medium is negligible initially, it will remain small at all times. Furthermore, stars take birth in this medium and we would predict negligible helium in stellar atmospheres. This is in contrast with actual observations. The interstellar medium contains about 28% helium by mass. We also observe that the relative abundance of ^{4}He seen in the first generation, Population II, stars is about

25% by mass. The helium abundance of Population I stars is also similar to this. If stars produced significant amounts of helium, then Population I stars would have been born in a medium with a higher proportion of this element, thus leading to higher helium abundance. The present best estimates suggest that primordial helium abundance lies in the range 23% to 25%. We conclude that this helium could have only been produced in the early Universe and not in stellar interiors.

The abundance of other light nuclei such as D, ^{3}He, ^{7}Li is very small. However, even this small abundance cannot be explained by stellar nucleosynthesis. For example, all the deuteron (D), which is produced in stars by hydrogen fusion, quickly gets consumed to form ^{3}He. The deuteron abundance in the interstellar medium is found to be D/H= 1.6×10^{-5}. One can also measure the primordial abundance of D from the absorption lines formed by clouds at high redshift in the spectra of distant quasars. This leads to a somewhat higher value, D/H= 3×10^{-5}. The Big Bang cosmology has been very successful in explaining the relative abundance of D and other light elements.

Let us now deduce the abundance of light elements produced in the early Universe. Here we will focus primarily on helium. At sufficiently early time, the density of helium and heavier nuclei was very small in comparison to that of protons and neutrons. Their density becomes significant only after the temperature is such that $kT < B$, where B is their binding energy. The first nuclei to form are deuteron (D). Heavier nuclei can form only after the formation of D. Hence this acts as a strong bottleneck, just as in the case of stellar nucleosynthesis. The binding energy of D is 2.22 MeV, which is very small. By the time this nucleus forms in substantial quantities, the medium has cooled so much that only a few light nuclei can form. Recall that the formation of nuclei heavier than helium proceeds by the triple alpha process. Due to the strong Coulomb barrier, this turns out to be insignificant in the early Universe. Furthermore, the time available for nucleosynthesis is very short due to the expansion of the Universe. The primordial nucleosynthesis actually starts only when kT is much smaller than the binding energy of D. This is because of the large value of η, Equation 16.57, which implies that the number density of photons is much larger than that of protons and neutrons. The high density of photons with energies of order 1 MeV would quickly break any D nuclei that may form, thus delaying the onset of nucleosynthesis.

We next determine the equilibrium distribution of protons, neutrons, and nuclei. These particles are nonrelativistic at the onset of nucleosynthesis. In this limit, $E/kT >> 1$, both the Fermi–Dirac and Bose–Einstein distributions reduce to the Boltzmann distribution. We approximate

$$E \approx mc^2 + \frac{p^2}{2m}.$$ (16.58)

The corresponding number density is given by

$$n = \frac{g_{int}}{2\pi^2\hbar^3} e^{(\mu - mc^2)/kT} \int_0^\infty p^2 dp \exp[-p^2/2mkT],$$ (16.59)

where μ is the chemical potential. This leads to

$$n = g_{int}\left(\frac{mkT}{2\pi\hbar^2}\right)^{3/2} e^{(\mu-mc^2)/kT} . \tag{16.60}$$

The factor $g_{int} = 2$ for protons and neutrons and 3 for D. Let n_p, n_n, and n_D denote the number densities of proton, neutron, and D, respectively, and μ_p, μ_n and μ_D their chemical potentials. The dominant reaction that contributes to the formation of D in the early Universe is

$$p + n \leftrightarrow D + \gamma . \tag{16.61}$$

At equilibrium, $\mu_D = \mu_n + \mu_p$, because the chemical potential of photons is zero. We have

$$n_p = 2\left(\frac{m_b kT}{2\pi\hbar^2}\right)^{3/2} e^{(\mu_p - m_p c^2)/kT} ,$$

$$n_n = 2\left(\frac{m_b kT}{2\pi\hbar^2}\right)^{3/2} e^{(\mu_n - m_n c^2)/kT} ,$$

$$n_D = 3\left(\frac{2m_b kT}{2\pi\hbar^2}\right)^{3/2} e^{(\mu_D - m_D c^2)/kT} , \tag{16.62}$$

where we have approximated $m_b \approx m_p \approx m_n$ and $m_D \approx 2m_b$ in the coefficient but not in the exponent. Eliminating μ_D, we obtain

$$n_D = \frac{3}{4}2^{3/2} n_n n_p \left(\frac{2\pi\hbar^2}{m_b kT}\right)^{3/2} e^{B/kT} , \tag{16.63}$$

where $B = (m_n + m_p)c^2 - m_D c^2$ is the binding energy of D. We next express this in terms of the mass fractions $X_D = 2n_D/n_b$, $X_p = n_p/n_b$, and $X_n = n_n/n_b$. Here the mass of D is approximately $2m_H$ and that of a proton and neutron approximately equal to m_H, where m_H is the mass of the hydrogen atom. Notice that when protons and neutrons are in equilibrium, $n_b \approx n_p + n_n$ and $X_n \approx X_p \approx 1/2$. We obtain

$$X_D = \frac{24}{\sqrt{\pi}}\zeta(3) X_n X_p \eta \left(\frac{kT}{m_b c^2}\right)^{3/2} e^{B/kT} . \tag{16.64}$$

In Figure 16.5 we plot the ratio $X_D/(X_p X_n)$ as a function of kT. For the time period corresponding to this plot, X_p and X_n are both of order unity. We find that the ratio is very small for $kT > 0.1$ MeV. It becomes close to unity only at $kT \sim 0.08$ MeV. This energy is much smaller than the binding energy of D. As explained earlier, the reason for the late onset of nucleosynthesis is the very small value of η. The helium formation starts once the deuteron fraction, $X_D \sim 0.1 - 1$. It happens over a very short time interval when 0.05 MeV $< kT <$ 0.09 MeV. The reactions contributing are

$$
\begin{aligned}
D + D &\rightarrow {}^3He + n \,, \\
{}^3He + D &\rightarrow {}^4He + p \,;
\end{aligned}
\tag{16.65}
$$

$$
\begin{aligned}
D + D &\rightarrow {}^3H + p \,, \\
{}^3H + D &\rightarrow {}^4He + n \,;
\end{aligned}
\tag{16.66}
$$

and the reactions involving photon emission

$$
\begin{aligned}
D + n &\rightarrow {}^3H + \gamma \,, \\
{}^3H + p &\rightarrow {}^4He + \gamma \,;
\end{aligned}
\tag{16.67}
$$

$$
\begin{aligned}
D + p &\rightarrow {}^3He + \gamma \,, \\
{}^3He + n &\rightarrow {}^4He + \gamma \,.
\end{aligned}
\tag{16.68}
$$

These reactions produce predominantly ^4He and small quantities of ^3H and ^3He. In order to compute the ^4He abundance, we assume that all the neutrons present when $kT = 0.07$ MeV form ^4He.

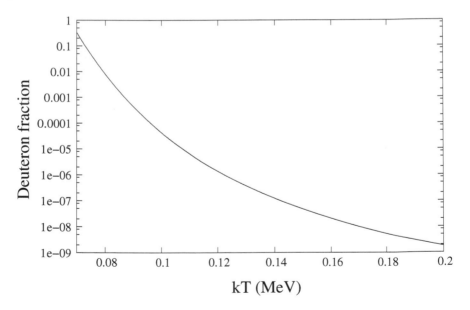

FIGURE 16.5: The deuteron fraction $X_D/(X_p X_n)$ as a function of kT during the early Universe. We find that the amount of deuteron is very small for $kT > 0.1$ MeV. Only at $kT < 0.1$ MeV does X_D becomes comparable to X_p. Hence nucleosynthesis occurs only below this temperature

The neutrons stay in equilibrium with the cosmic plasma until $kT \approx 0.75$ MeV. Their equilibrium with the medium is maintained by the processes

$$
\begin{aligned}
\nu + n &\leftrightarrow p + e^{-}, \\
e^{+} + n &\leftrightarrow p + \bar{\nu},
\end{aligned} \tag{16.69}
$$

and the neutron decay process

$$
n \rightarrow p + e^{-} + \bar{\nu}. \tag{16.70}
$$

These processes involve weak interactions. The neutrons decouple when their reaction rate, Γ, becomes smaller than the Hubble parameter, $H(t)$. At equilibrium, the ratio of neutron to proton number densities is equal to

$$
\frac{n_n}{n_p} = e^{(-m_n + m_p)c^2/kT}, \tag{16.71}
$$

where we have used $\mu_n \approx \mu_p$ assuming that $\mu_e \approx 0$ and $\mu_\nu \approx 0$. The difference $m_n c^2 - m_p c^2 \approx 1.29$ MeV. At the neutron decoupling temperature, $kT \approx 0.75$ MeV, we obtain, $n_n/n_p \approx 0.18$. This leads to $X_n \approx 0.15$. Beyond this time, neutrons propagate freely. The ratio n_n/n_p decreases during this free propagation due to the neutron decay process. The lifetime of a free neutron is approximately $t_n \approx 887$ sec. Using this we can compute the neutron abundance at $kT \approx 0.07$ MeV, the temperature corresponding to primordial nucleosynthesis.

The time elapsed between neutron decoupling and the onset of primordial nucleosynthesis can be computed using the cosmic evolution equations. Let us consider the evolution of primordial plasma starting from a temperature of about $10^{10} K$ ($kT \approx 1$ MeV). At this temperature, the species that are in equilibrium are photons, electrons, positrons, protons, and neutrons. The neutrinos have decoupled but are still relativistic. At temperatures of order 1 MeV, protons, neutrons, and nuclei are nonrelativistic. The energy density is dominated by the relativistic species and is given by

$$
\rho = \frac{g_{int}}{2\pi^2 \hbar^3} \int_0^\infty \frac{E p^2 dp}{\exp(E/kT) \pm 1}, \tag{16.72}
$$

where the $-$ and $+$ signs correspond to bosons and fermions respectively. Here we have neglected the chemical potentials, which is a good approximation for all the species. We obtain

$$
\begin{aligned}
\rho &= \frac{g_{int} \pi^2}{30} \frac{(kT)^4}{(\hbar c)^3} \quad \text{for bosons}, \\
\rho &= \frac{7}{8} \frac{g_{int} \pi^2}{30} \frac{(kT)^4}{(\hbar c)^3} \quad \text{for fermions}.
\end{aligned} \tag{16.73}
$$

The total energy density is obtained by summing over all the species. The factor g_{int} is equal to 2 for photons, electrons and positrons, and 1 for each of the three species of neutrinos and anti-neutrinos. We obtain

$$\rho = \frac{g_{eff}\pi^2}{30} \frac{(kT)^4}{(\hbar c)^3}\,, \tag{16.74}$$

where

$$g_{eff} = 2 + \frac{7}{8}(4+6) = 10.75 \tag{16.75}$$

sums over the contribution from all the species. Substituting this into Equation 16.24, we can obtain the relationship between time and temperature during this era. We recall that during the radiation domination phase, $a(t) \propto t^{1/2}$. Hence $H(t) = 1/(2t)$. Using Equation 16.24, we obtain

$$t = \sqrt{\frac{45}{16\pi^3 g_{eff}}} \sqrt{\frac{\hbar^3 c^5}{(kT)^4 G}}\,. \tag{16.76}$$

This leads to

$$t = 0.736 \text{ sec } \left(\frac{1 \text{ MeV}}{kT}\right)^2. \tag{16.77}$$

We also need this relationship at the time of nucleosynthesis, $kT \approx 0.07$ MeV. By this time the electron-positron pairs have annihilated. Hence now only the photons and neutrinos contribute significantly to the energy density. The neutrino temperature, Equation 16.49, is a little lower. We can absorb these changes in the g_{eff}, whose value now is given by, $g_{eff} = 3.36$. Correspondingly, the relationship between time and temperature becomes

$$t = 132 \text{ sec } \left(\frac{0.1 \text{ MeV}}{kT}\right)^2. \tag{16.78}$$

The time corresponding to $kT = 0.07$ MeV is 269 seconds. Taking into account neutron decay, we find that at this time, $X_n \approx 0.11$. Let Y_p denote the primordial ^{4}He mass fraction. Using Equation 8.45 with $M = n_b m_H$, $m_H \approx m_b$ and assuming that all the neutrons get converted to ^{4}He, we obtain

$$Y_p = 2X_n \approx 0.22\,. \tag{16.79}$$

A more detailed calculation gives a slightly larger value of about 0.24, which agrees with the observed value of Y_p which lies approximately in the range, 0.23 to 0.25.

16.5.2 Recombination

After the formation of light nuclei, the primordial plasma consists of photons, electrons, protons, and light nuclei. The protons and light nuclei remain in equilibrium with the plasma due to Coulomb interaction with electrons. The electrons maintain equilibrium due to Compton scattering with photons. As the temperature cools further, we reach the epoch in which the energy density of radiation becomes equal to that of matter. Beyond this time, the Universe is dominated by nonrelativistic matter. A little later, electrons start combining with protons and nuclei to form atoms. This is called recombination, although this happens to be the first time in the history of the Universe that the nuclei and electrons combine to form atoms. This process happens when $kT \sim 0.1$ eV, which is much smaller than the binding energy of atoms. This delay happens for precisely the same reason as in the case of primordial nucleosynthesis. The extremely large number density of photons does not allow atoms to form at higher temperatures. The redshift corresponding to this event is estimated to be about 1,000. Due to recombination, the free electron density becomes very small. Hence the Compton scattering rate, which kept the free electrons in equilibrium with photons, drops below the Hubble expansion rate. The photons decouple from electrons and propagate freely in space. The medium becomes transparent to photons. The resulting photon gas is observed today as Cosmic Microwave Background Radiation (CMBR).

After recombination, the Universe entered the era that is called the dark ages. During this time, the Universe was filled with neutral light atoms. Hence it is very difficult to detect any signal from this era. The only possibility is the 21 cm line due to hyperfine splitting of neutral hydrogen. Due to the expansion of the Universe, this line is redshifted to wavelengths on the order of several hundred centimeters. There is currently considerable effort to devise methods to observe this signal.

The dark ages ended after the formation of structures in the Universe. As the first galaxies appeared, ultraviolet light from the stars ionized the neutral atoms. This phenomenon is called reionization. The current best estimates suggest that reionization occurred at $z \approx 6$.

16.5.3 Structure Formation

The structures, that is, galaxies, clusters, etc., form due to perturbations around the uniform background density. We have so far confined ourselves to perfectly homogeneous and isotropic space-time. The density of matter as well as the curvature of space-time were assumed to be same at all points and directions. However, there are fluctuations or perturbations around this smooth background. These were generated during a very early phase of the Universe called inflation. These perturbations evolved with time but stayed relatively small until the end of the radiation dominated phase. The density $\rho(\vec{x})$ of matter at position $\vec{x}$, can be expressed as

$$\rho(\vec{x}) = \rho_0 + \delta\rho(\vec{x}), \qquad (16.80)$$

where ρ_0 represents the smooth mean density, which is independent of position, and $\delta\rho(\vec{x})$ represents the perturbations. The gravitational attraction causes the perturbations to grow, whereas the outward pressure causes them to decay. This is the same as the growth of perturbations in a cloud of gas and dust during star formation. The gravitational attraction starts to dominate during the nonrelativistic matter dominated phase. The perturbations grow to eventually form structures in the Universe. The theory of structure formation predicts that

$$\frac{\delta\rho}{\rho_0} \sim 10^{-5} \tag{16.81}$$

at the time when photons decoupled from matter.

16.5.4 Cosmic Microwave Background Radiation (CMBR)

As described above, photons decouple from electrons at $z \approx 1,000$, a little after recombination. At the time of decoupling, these photons had a blackbody spectrum with $kT \approx 0.1$ eV. Subsequently, the photons propagate freely with their blackbody temperature decreasing as $1/a$. This essentially means that their wavelengths are redshifted due to expansion. This radiation is observed today at microwave frequencies and called CMBR. It spectrum matches the blackbody distribution with $T = 2.73$K, with very high precision. The temperature is observed to be almost the same in all directions. This provides considerable justification for our assumption that the Universe is isotropic.

The CMBR temperature is not exactly isotropic. The largest deviation from isotropy arises due to the peculiar velocity of the Milky Way. It is moving in some direction due to the gravitational field of the galaxies in our neighborhood. Furthermore, the solar system is in motion with respect to the Milky Way center. Hence the observed photon frequencies get shifted due to the Doppler effect. This leads to a dipole pattern in the CMBR temperature. In galactic coordinates, the direction of the dipole is found to be $l = 264°$ and $b = 48°$. Choosing this to be the z-axis, the resulting CMBR temperature can be written as

$$T(\theta, \phi) = T_0 + \delta T_d \cos\theta, \tag{16.82}$$

where $T_0 = 2.73$K is the mean temperature, θ the polar angle, and δT_d represents the strength of the dipole. Let v represent the speed of the solar system with respect to the frame in which the CMBR is isotropic. Then

$$\frac{\delta T_d}{T} \approx \frac{v}{c} \sim 10^{-3}. \tag{16.83}$$

The observed value of $v = 369$ Km/s.

Besides the dipole pattern, CMBR displays anisotropies that arise due to primordial density fluctuations. The energy density of the photon gas is also not exactly uniform. Its perturbations are described by an equation analogous to Equation 16.80. As long as the photons remained in equilibrium with electrons, the photon fluctuations, $\delta\rho/\rho_0$, were of the same order as those of the matter field. Because the energy density of photons is proportional to T^4, this implies that $\delta T(\vec{x})/T_0 \sim 10^{-5}$. Here T_0 is the mean temperature and δT

represents the fluctuations. After decoupling, the photons propagate freely. Hence the strength of their perturbations did not change, in contrast to those of matter perturbations. We observe these photons from our location. Because the photon gas was inhomogeneous at the time of decoupling, the photons that we observe from different directions would have a slightly different temperature. Hence we expect to see a small anisotropy in their temperature. We can, therefore, express the observed temperature of CMBR as

$$T(\theta, \phi) = T_0 + \delta T(\theta, \phi), \tag{16.84}$$

where $\delta T(\theta, \phi)$ is a function of the angular coordinates θ and ϕ. Remarkably these fluctuations were observed close to the predicted value of $\delta T/T_0 \sim 10^{-5}$ by the COBE satellite in early 1990s. The CMBR anisotropies have now been studied in great detail by WMAP and PLANCK satellites. Their observations provide very accurate values of the cosmological parameters and find good agreement with the Lambda Cold Dark Matter (LCDM) model.

This concludes our discussion on cosmology. A more detailed treatment requires the mathematical structure of the general theory of relativity, which is beyond the scope of this book.

Exercises

16.1 In this problem we develop a model of expansion based on Newtonian mechanics. Assume that the Universe is filled with nonrelativistic matter of energy density ρ. The gravitational force on a test mass m (see Figure 16.6), located at a distance s from origin, is given by

$$m\ddot{s} = -\frac{GM(s)m}{s^2},$$

where $M(s)$ is the total mass contained within a volume of radius s, and the negative sign indicates that the force is attractive. Notice that by spherical symmetry, the mass outside radius s does not contribute.

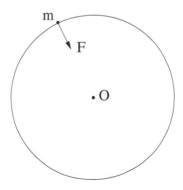

FIGURE 16.6: The force $\vec{F}$ on a test mass m located at distance s from the origin O due to a uniform mass distribution in space.

(a) Set $s = a(t)r$, where r is fixed and $a(t)$ acts as the scale parameter. Determine the differential equation for $a(t)$.

(b) Verify that for nonrelativistic matter this equation is equivalent to the set of Equations 16.24 and 16.25. How does ρ depend on a?

16.2 The wavelengths of some visible lines in helium (HeI), calcium (CaII), and hydrogen (HI) spectra, as observed in laboratory, are 381.9 nm, 393.3 nm, 396.8 nm, 400.9 nm, and 410.2 nm. The wavelengths of some of the spectral lines we observe from a distant galaxy are 511.3 nm, 515.8 nm, and 533.3 nm. Match these lines with those observed in the laboratory and deduce the redshift of this galaxy.

16.3 Numerically verify that the value of the critical energy density is given by Equation 16.27.

16.4 Verify that for nonrelativistic matter, relativistic matter, and the cosmological constant, the scale parameter is given by Equations 16.28, 16.28, and 16.44, respectively.

16.5 Obtain the equation for co-moving distance, Equation 16.37, starting with Equation 16.33.

16.6 Obtain the formula for luminosity distance d_L for a Universe dominated by (a) nonrelativistic matter and (b) a cosmological constant.

16.7 Verify Equation 16.46, which gives the lifetime of the Universe within the framework of the LCDM model. Numerically determine the age assuming that $\Omega_M \approx 0$.

16.8 Verify that for $w < -1/3$, the evolution equations, Equations 16.24 and 16.25, lead to accelerating Universe $\ddot{a} > 0$.

16.9 Obtain Equation 16.56 by performing the integral in Equation 16.55. Use the fact that for photons, $E = pc$. You can use the integral

$$\zeta(x) = \frac{1}{\Gamma(x)} \int_0^\infty \frac{s^{x-1}}{e^s - 1} ds,$$

where $\Gamma(x)$ is the standard Gamma function ($\Gamma(3) = 2$).

16.10 Verify that the numerical value of the parameter η defined in Equation 16.52, is given by Equation 16.57.

16.11 Verify Equation 16.64 for the mass fraction of deuteron.

16.12 Obtain the time-temperature relationship, Equation 16.76, during the radiation dominated era.

16.13 Obtain the relationship, $Y_p = 2X_n$, for the mass fraction of helium produced during primordial nucleosynthesis.

Chapter 17

Active Galaxies

17.1   Introduction

Active galaxies are an exotic class of galaxies, whose properties are very different in comparison to normal galaxies. Their most striking feature is their nuclei, which is very small and extremely luminous. In some cases, the size of these nuclei may be comparable to that of the solar system. However, their luminosity is typically larger than that of a whole galaxy. Hence these cores generate a huge amount of power in a very compact region and are called Active Galactic Nuclei or AGNs. The host galaxies of AGNs may be one of the normal galaxies, such as a spiral, elliptical, etc. These hosts are often difficult to observe because they are outshined by their nuclei.

The AGNs are so compact that it has not been possible to resolve them, that is, we are unable to directly measure their size. However, their luminosity shows a large variation over a relatively short time interval that allows us to impose an upper limit on their size. Furthermore, we can also estimate a lower limit on their mass. Such analysis suggests the existence of a very massive black hole, with mass of order 10^8 M_{Sun} at the center of these objects.

A striking feature of these nuclei is that they emit radiation in a wide range of frequencies, such as, radio, infrared, visible, ultraviolet, x-rays, and γ-rays. In many cases, the nucleus appears almost like a star at visible wavelengths. Its true nature becomes clear only when one determines its luminosity, spectral lines, as well as continuum emission at different wavelengths. The spectrum of a normal galaxy is dominated by absorption lines, produced in stellar atmospheres. In contrast, the AGNs display a large number of emission lines. These can only be produced by atoms and ions in excited states. Hence the AGN temperature must be very high. In many cases we observe rather broad emission lines. This broadening can be attributed to Doppler shifts caused by the large velocities of these ions or atoms in random directions due to high temperature in AGN atmospheres.

Active galaxies are found to be more abundant at large distances in comparison to our local neighborhood. It is useful to keep in mind that when we observe the Universe at large distances from Earth we get a glimpse into its structure at early times. This is due to the time that light takes to reach us. Hence if we observe an object at a distance of 1 Gpc, light that we observe left the object roughly 3.26 billion years ago. This suggests that in the past, active galaxies dominated the Universe and for some reason have largely disappeared in recent times.

17.2 Active Galactic Nuclei: Some Basic Properties

AGNs display a wide range of bewildering phenomena, that have fascinated astrophysicists for many decades. These objects are extremely luminous, with their luminosity often surpassing that of a regular galaxy by many orders of magnitude. This is particularly amazing, given the fact that their size is very small. Furthermore, they have significant continuum emission over a wide range of frequencies from radio waves to γ-rays. They display both emission and absorption spectral lines. Their luminosity also shows a rapid variation with time.

All these properties indicate the presence of a very compact and highly active object. The compact nature of this object is deduced from the rapid time variability. Furthermore, one deduces the mass of this object from its luminosity. We discuss these details later in this chapter. Here we describe the basic picture of an AGN that is consistent with observations. The luminosity produced by this compact object is so large that even nuclear fusion does not have the efficiency required to generate such high power. The only possible mechanism is the gravitational attraction of a very massive black hole. The mass of the black hole is found to lie in the range 10^6 to 10^{10} solar masses. The black hole attracts matter from its neighborhood, which forms an accretion disk, as illustrated in Figure 17.1. The figure schematically illustrates a central black hole, surrounded by an accretion disk, both of which may be undergoing a slow spin. The system emits jets of charged particles, protons and electrons, in two opposing directions.

The formation of an accretion disk is a phenomenon similar to that discussed in the case of a binary system, Figure 10.6. However, in that case a binary partner feeds matter into a black hole or a very compact object, with mass on the order of a solar mass. In the present case, the black hole is very massive. Furthermore, it may be accreting cold matter from all directions. The matter attracted by the black hole has to pass through the accretion disk. As it collides with matter in the disk, it generates heat and radiation. This leads to the extreme luminosity of AGNs in infrared, visible, ultraviolet, x-rays and γ-rays. Radiation at these frequencies comes directly from the central compact region of an AGN.

AGNs also emit jets of charged particles that travel very large distances. The precise mechanism for the formation of these jets is unknown. It is possible that charged particles may be accelerated to very high energies in an AGN, perhaps due to time-varying magnetic and electric fields. These fields may be generated by rotation of the AGN. For example, a time-dependent magnetic field is generated if the magnetic dipole axis does not align with the rotation axis. A time-varying magnetic field also creates a time-varying electric field. These energetic charged particles can be channeled into jets by magnetic field

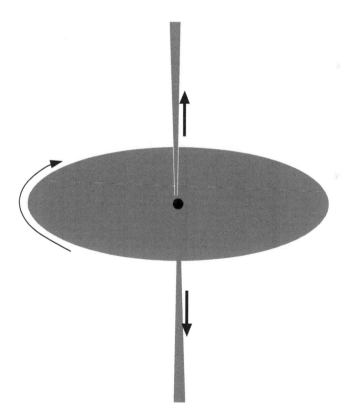

FIGURE 17.1: Here we schematically show an AGN with a central black hole and an accretion disk. Two jets of charged particles are emitted in opposite directions.

lines, much like the solar wind, shown in Figure 11.6. We point out that the phenomenon of jet formation is also seen in the case of young stars, as discussed in Chapter 10.

The high level of activity of an AGN is possible only if a sufficiently high density of cold matter is available in the vicinity of the black hole. Such conditions might have arisen when the galaxies first formed out of a primordial medium. As discussed in the previous chapter, galaxies form due to fluctuations in matter density. Regions of large density attract matter and grow. The central region of these dense structures may eventually evolve into a black hole. The density of matter surrounding the black hole at this early stage may be sufficiently high to sustain the level of activity seen in AGNs. The matter slowly depletes, and eventually the activity reduces to the level seen in normal galaxies. Hence we expect that active galaxies are the early stages of galaxy formation. All active galaxies may slowly evolve into normal galaxies, with a dormant, massive black hole at their centers.

Active galaxies might also arise by a collision between two normal galaxies. This may lead to a compression and hence accumulation of gas and dust near the center of the system. The dormant black hole near the center of one, or both the galaxies might again become active and start behaving as an AGN. We expect that in this case the structure of the galaxy will be considerably distorted, which is observed in the case of many AGNs.

Due to the presence of the disk, these objects do not have spherical symmetry. They may be symmetric about an axis perpendicular to the accretion disk. Hence their emission is not expected to be uniform in all direction and their appearance depends on the angle of observation. This is similar to what is seen in the case of, for example, spiral galaxies. However, in the present case, the dependence on the angle of observation appears to be very dramatic, as we shall discuss later.

17.2.1   Size of AGNs

As mentioned above, AGNs show very rapid fluctuations in luminosity. This indicates that their size is very small. A significant change in luminosity can occur only if the change happens systematically throughout the entire source. A source may have local random fluctuations in emission. When summed over the entire source, such changes will add up to zero. Consider a source of length D whose luminosity undergoes a significant change, on the order of 50%, in time Δt. The information that this change has occurred propagates along the source at speed less than c, the speed of light. Hence the information reaches the opposite end in time greater than D/c. This provides an estimate of the minimum time over which the luminosity can undergo significant change. This implies that $\Delta t > D/c$, or equivalently,

$$D < c\Delta t. \tag{17.1}$$

This gives an upper limit on its size. The time interval Δt for AGNs is observationally found to be in the range of an hour to about a month. In Exercise 17.2 you can show that the size corresponding to 1 hour turns out to be smaller than that of the solar system.

17.2.2 Luminosity

The AGNs are extremely luminous. Their large luminosity allows us to impose a lower bound on their mass. Consider an object of mass M. There has to be a limit on the luminosity that it can produce, while maintaining equilibrium. This is because large luminosity means large radiation flux, which would lead to large radiation pressure. If the radiation pressure becomes too large, it might overcome the attractive forces that hold the object together and blow it apart. The limiting value of luminosity is called the Eddington limit. It is given by

$$L_{max} = \frac{4\pi cGM}{\bar{\kappa}}, \qquad (17.2)$$

where $\bar{\kappa}$ is the Rosseland mean opacity, discussed in Section 8.6. We obtain this limit by considering a spherically symmetric object of mass M. The limit is reached when the radiation pressure near the surface, at radius r, balances the gravitational attraction. The radiation pressure P and temperature T are related by Equation 8.54. Using this along with the equation for radiative transport, Equation 8.25, we obtain

$$\frac{dP}{dr} = -\frac{\bar{\kappa}\rho L}{4\pi cr^2}, \qquad (17.3)$$

where ρ is the density of the medium. Assuming that the dominant attractive force acting on the object is the gravitational force, the pressure gradient is given by Equation 8.7. We deduce from Equations 8.7 and 17.3 that the maximum luminosity for an object of mass M is given by Equation 17.2. The details of this derivation are left as an exercise. If the luminosity exceeds this value, then the pressure gradient due to radiation pressure is too large to be sustained by gravitational attraction and the object will not remain in equilibrium. Here we have made several assumptions, such as spherical symmetry. However, Equation 17.2 provides a reliable order of magnitude estimate in most cases.

At very high temperatures, the medium gets completely ionized and the dominant contribution to $\bar{\kappa}$ comes from Compton scattering (see Section 8.6). Using the definition of opacity, Equation 7.11, we find that $\kappa_\lambda = n\sigma_\lambda/\rho$, where n is the number density. The scattering cross section is given by the Thomson cross section σ_T, given in Equation 8.31, which is independent of wavelength. Furthermore, we assume that the medium is predominantly ionized hydrogen, which leads to $\rho/n \approx m_H \approx m_p$. In this case we can express the Eddington luminosity as

$$L_{max} = \frac{4\pi cGMm_p}{\sigma_T}$$

$$\approx 1.26 \times 10^{31} \left[\frac{M}{M_{Sun}}\right] \text{ W} \approx 3.3 \times 10^4 L_{Sun} \left[\frac{M}{M_{Sun}}\right]. \quad (17.4)$$

Hence we find that for a solar mass object, the maximum luminosity it can produce, in equilibrium, is on the order of 10^4 times the luminosity of the Sun. Let us now apply this to the AGN 3C273. It has a luminosity on the order of $3 \times 10^{12} \ L_{Sun}$. This implies that its mass must be larger than 10^8 solar masses.

17.2.3 Superluminal Motion

Some of the AGNs show superluminal motion, that is, a part of the object appears to be moving away at a speed greater than the speed of light. What we measure is the transverse speed of the object on the sky. How can this speed exceed the speed of light? Does this violate special relativity? The situation is explained in Figure 17.2. Consider an AGN located at A, at a distance D from an observer. Assume that a segment moves away from it toward B at speed v at an angle θ, as shown in Figure 17.2. Light signal from A, emitted at time $t = 0$, reaches the observer at time $t_1 = D/c$. The segment reaches B after a time interval equal to L/v. Let the light signal from this segment at B reach the observer at time t_2. Using the geometry shown in Figure 17.2, you can show as an exercise (Exercise 17.4) that

$$\Delta t = t_2 - t_1 = \frac{L}{v} - \frac{L\cos\theta}{c}. \quad (17.5)$$

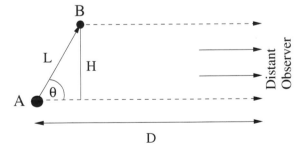

FIGURE 17.2: A segment of an AGN is seen to move from position A to B in the sky in time Δt, at an angle θ with respect to the line of sight. The object travels transverse distance H, given by, $H = L\cos\theta$, where L is the total distance traveled. (Adapted from D. J. Griffiths, *Introduction to Electrodynamics*.)

According to the observer, the segment travels an apparent distance equal to H in this time interval. Hence its apparent speed u is

$$u = \frac{H}{\Delta t} = \frac{v \sin \theta}{1 - \frac{v}{c} \cos \theta}. \tag{17.6}$$

As an exercise (Exercise 17.5), you can show that the apparent speed is maximum when

$$\cos \theta_{max} = v/c. \tag{17.7}$$

The corresponding maximum apparent speed in this case is

$$u_{max} = \frac{v}{\sqrt{1 - v^2/c^2}}. \tag{17.8}$$

Hence we find that for v close to c, the apparent speed can far exceed the speed of light. This happens if θ is very close to zero, that is, the motion is in a direction close to the line of sight. Hence it is only the apparent speed that is greater than c and there is no violation of the principle of special relativity.

17.3 Classification of Active Galaxies

The active galaxies are broadly classified as Seyfert galaxies, blazars, radio galaxies, and quasars. These objects differ in their emission characteristics, that is, the nature of spectral lines, their overall luminosity, as well continuum emission at different wavelengths.

17.3.1 Seyfert Galaxies

Most of the Seyfert galaxies are spirals with a core consisting of an AGN. Approximately 2% of all spiral galaxies are Seyferts. The peculiar nature of their nuclei was first observed by Carl K. Seyfert in 1943. At visible wavelengths, these galaxies appear as normal spiral galaxies. However, when viewed at other wavelengths, we observe a very luminous, unresolved nucleus. These galaxies are further classified on the basis of their spectral lines into Seyfert 1 (Sy 1) and Seyfert 2 (Sy 2), along with intermediate types such as Seyfert 1.5. Seyfert 1 galaxies show both broad and narrow emission lines. Assuming that these lines are broadened due to the Doppler effect, we estimate velocities of gas particles in the range of 1,000 to 5,000 Km/s in these objects. These galaxies show strong emission in ultraviolet and x-rays. Their x-ray emission is also highly variable. In contrast, Seyfert 2 galaxies show only narrow emission lines, which indicate velocities of order 500 to 1,000 Km/s. Furthermore, these galaxies show strong infrared emission and may not have high luminosity in x-rays or ultraviolet.

Up to now, about 15,000 Seyfert galaxies have been observed. An example is NGC 5548, also called MrK1509, located at a distance of about 75 Mpc ($z = 0.017$), in constellation Böotes. It is classified as Seyfert 1.5. The nucleus shows highly variable emission, both at visible and x-ray frequencies. An example of Sy 2 is NGC 1068 (Messier 77). It is located at a distance of about 14 Mpc in constellation Cetus.

17.3.2   Radio Galaxies

Some galaxies have very strong radio emissions, much larger in comparison to that of normal galaxies. A radio image of these sources typically displays two widely separated lobes, which may be connected by jets to a central region. It is believed that the radio emission is powered by an active nucleus that resides in the core of these galaxies. These objects are called radio galaxies. An example is Cygnus A, located at a distance of about 230 Mpc ($z = 0.056$) in constellation Cygnus. A radio image of this source is shown in Figure 17.3. The narrow jets consist of high energy electrons emitted by the core in opposite directions and moving at speeds close to the speed of light. These electrons most likely originate in the active nuclear region of these galaxies and are confined to narrow jets by a dipolar magnetic field, aligned parallel to the jet axis. At large distance from the center, these electrons collide with the intergalactic medium and generate very strong radio emission. These regions are identified with the radio lobes, which typically dominate the radio emission of these objects.

In most cases the host galaxies of these sources are elliptical. Hence if we view them at visible frequencies, we observe an elliptical galaxy. In many cases the host galaxy appears highly distorted. This is most likely due to the effect of other galaxies or due to a merger. The radio component of these objects is typically much larger than the host galaxy. For example, the two radio lobes of

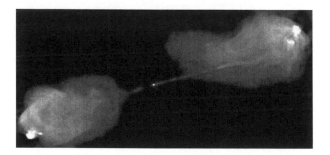

FIGURE 17.3: The radio galaxy, Cygnus A, imaged at radio frequencies (5 GHz). We observe two very bright lobes connected to a point-like core with two jets. The radio emission is brightest at the extreme ends of the lobes where they meet the intergalactic medium. (Image courtesy of NRAO/AUI.)

Cygnus A extend well beyond the host galaxy into the intergalactic medium. The distance between the two hot spots, that is, regions of strongest emission in Figure 17.3, is of the order of 100 Kpc, three times larger than the disk diameter of the Milky Way.

These sources show a continuum, nonthermal spectrum, characteristic of synchrotron emission due to high-energy electrons accelerating in a magnetic field. The spectrum of can be well described by a power law,

$$F(\nu) \propto \frac{1}{\nu^\alpha}, \qquad (17.9)$$

where $F(\nu)$ is the observed flux at frequency ν and α the spectral index. It is often possible to fit the spectrum with a constant value of $\alpha \approx 1$ over a limited range of frequencies. Over a larger range, α shows dependence on ν. The observed spectrum arises due to the integrated emission from a large number of electrons.

The observed radio flux of Cygnus A at 14.7 MHz is approximately 31,700 Jy. The spectral index $\alpha \approx 0.7$ in the range 10 MHz to 2 GHz. Below 10 MHz, the flux cuts off sharply and above 2 GHz, the spectral index becomes larger, $\alpha \approx 1.2$. As an exercise (Exercise 17.1), you can use this information to show that its total luminosity in this frequency range is on the order of 10^{44} ergs/s. This is many orders of magnitude larger than the radio luminosity of normal galaxies and comparable to the total luminosity of the Milky Way, integrated over all wavelengths.

The strongest radio emission comes from the hot spots in the lobes. These hot spots are typically seen where the jet of particles meets the intergalactic medium. This phenomenon may be understood in analogy with the interaction of the solar wind with the Earth's magnetic field, Figure 11.6. The magnetic field in this region is enhanced due to the compression generated by the collision of jet particles with the intergalactic medium. Furthermore, charged particles cruising nearly parallel to magnetic field lines suddenly encounter transverse magnetic fields and get trapped, leading to enhanced synchrotron emission.

In Cygnus A, we observe that one of the jets is much brighter than the other. In many cases, only one jet is visible and, in some, the jets may not be seen at all. This can be understood by recalling that synchrotron emission is dominant in the direction of motion of electrons. The velocity vector of electrons is nearly parallel to the jet axis. However, they do have a small velocity component perpendicular to the jet magnetic field, which makes them spiral around the jet axis. These spiraling electrons emit synchrotron radiation predominantly along the jet axis. The strength of emission depends on the angle between the jet axis and the line of observation. The strongest emission is seen if the jet is pointed directly toward us. This probably explains the strong emission from blazars, to be discussed later. In the case of radio galaxies, we may not observe radio waves from jets if the jet axis is aligned at a large angle to the line of sight.

17.3.3   Quasars

Quasars have very high luminosity, which is typically much larger than that of a normal galaxy. They are also the most distant objects seen in the Universe. Quasars have been observed in the redshift range $z =0.056$–7.085, and are the most abundant objects at high redshifts. The closest quasar lies at a distance of about 200 Mpc. Most are found at distances of order Gpc. More than 100,000 quasars have been observed so far. Their host galaxy is typically an elliptical galaxy. At Gpc distances, we do not observe many normal galaxies. This may be partially due to their low luminosity. Furthermore, the fact that quasars are seen abundantly at high redshifts and none in our vicinity suggests that they might represent an early stage of galaxy formation.

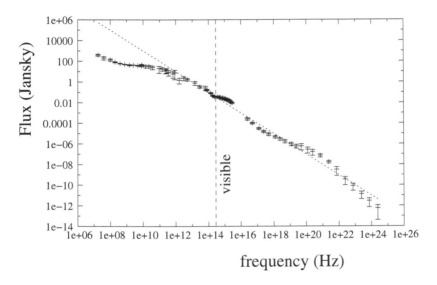

FIGURE 17.4: The spectrum of quasar, 3C273, from radio waves to gamma rays. The vertical, long dashed line shows the position of the visible spectrum. The slanted, short dashed line corresponds to a spectral index 1, that is, $1/\nu$, where ν is the frequency. The data for this plot have been taken from M. Türler, S. Paltani, and T. J.-L Courvoisier et al., Astronomy and Astrophysics Supplement, 134, 89 (1999) and S. Soldi, M. Türler and S. Paltani et al., Astronomy and Astrophysics, 486, 411 (2008).

At optical frequencies, quasars display a star-like appearance. However, their spectra are very different from the blackbody spectrum of a star. The spectrum of a typical quasar is relatively flat and extends over a wide range of frequencies. In Figure 17.4 we show the spectrum of the quasar, 3C273, located in the constellation Virgo. We observe a relatively flat spectrum at radio frequencies. At higher frequencies, we find a power law decay, Equation 17.9, with spectral index $\alpha = 1$. Due to their star-like appearance, these objects are called Quasi Stellar Objects (QSOs), or simply quasars. About 10% of all quasars have a radio galaxy associated with them. These are called radio-loud. The remaining are called radio-quiet.

The luminosity of quasars shows a very rapid variation in time. It might change quite significantly within a time interval of a few days to a few months. This imposes an upper limit on their size on the order of a few light days to a few light months.

The quasar spectrum displays broad emission lines. In this respect they are similar to Seyfert 1 objects. They also display absorption lines that are partially intrinsic to the quasar and partially arise during propagation through the intergalactic medium. The intergalactic medium at high redshifts or large distances contains a large number of clouds, which have a high density of neutral hydrogen, as shown in Figure 17.5. The quasar radiation that we observe at Earth has to pass through these clouds. These clouds produce Lyman α absorption line corresponding to the transition from $n = 1$ to $n = 2$ states of neutral hydrogen. A cloud at redshift z_i will produce a line at wavelength

$$\lambda_i = \lambda_\alpha(1 + z_i), \tag{17.10}$$

where λ_α is the wavelength of the Lyman α transition in the rest frame of the atom. Because there exist many clouds at different redshifts, we would observe many absorption lines corresponding to the different redshifts of different clouds. This dense collection of observed lines is called the Lyman α forest. It is of great importance in cosmology because it allows one to study the intergalactic medium at high redshifts.

FIGURE 17.5: The radiation emitted by a distant quasar, Q, passes through many clouds of neutral hydrogen before reaching the observer O. Here the clouds are shown at redshifts z_1, z_2, and z_3. Each cloud produces an absorption line whose wavelength is given by Equation 17.10, leading to a Lyman alpha forest in the observed quasar spectrum.

A very bright and well-studied quasar is 3C273. We have already described its spectrum in Figure 17.4. Its redshift is 0.158. This implies that it is located at a distance of about 700 Mpc, which, by quasar standards, is relatively close to us. The apparent visual magnitude V of 3C273 is 12.8. As an exercise, you can show (Exercise 17.3) that its visual luminosity is about $3 \times 10^{12}\ L_{Sun}$ or about 100 times the luminosity of the Milky Way. This is remarkable, given that its size is of the order of a few light months, much smaller than a parsec.

17.3.4   Blazars

Blazars show the strongest activity among the AGNs. All blazars have strong radio emission. Their emission at all wavelengths shows very rapid variation. In some cases, the time scale of variation is as small as a few hours. They also show superluminal motion, very high luminosity, and high polarization, both at radio and optical frequencies. These observations suggest that blazars might be AGNs, whose jets are pointed close to the line of sight. The jets contain charged particles moving at relativistic velocities. Due to relativistic boosting, their synchrotron emission is strongest in the forward direction. This also explains the superluminal motion seen in blazars. A lump of dense plasma, which is emitted along the jet pointing toward the observer, will have an apparent speed much larger than the speed of light, as discussed in section 17.2.3. The rapid time dependence of blazars indicates that the emission comes from a very compact object, whose size may be on the order of a few light hours.

17.4   Unified Model of AGNs

The AGNs display a diverse range of properties, as described above. It is believed that the main features of an AGN are a central black hole surrounded by an accretion disk, as illustrated schematically in Figure 17.6. The accretion disk is surrounded by a thick torus that lies in the same plane as the disk. Due to this torus, the central object is not visible to an observer whose line of sight lies in this plane. Furthermore, the AGN emits two jets of highly energetic charged particles perpendicular to this plane. These jets travel very large distances. The accretion disk as well as the obscuring torus are much smaller than the host galaxy; however, the jets are typically much larger. As described above, the jets terminate in the intergalactic medium, where they emit very strong radio waves in a lobe-like structure. The resulting structure appears as a radio galaxy.

If we view the object at higher frequencies, we can observe the central compact region. In particular, if the line of sight is inclined at an angle relative to the plane of the accretion disk, it looks like a quasar or Seyfert I object. We point out that quasars and Seyfert 1 objects are somewhat similar in character in the sense that they are both very bright at optical frequencies and display broad spectral lines. The main difference between the two is that quasars are much more powerful. The broad spectral lines result due to emission from the hot medium surrounding the central region, labeled in Figure 17.6 as Broad Line Region (BLR). If the line of sight lies close to the plane of the torus, then the central region is obscured from view. In this case we can only see the narrow spectral lines formed by cooler medium at larger distances from the center and the AGN appears as Seyfert 2 object. Finally the AGN appears as a blazar if viewed close to the jet axis. Hence the basic picture, illustrated in Figure 17.6, provides a unified picture of all AGNs.

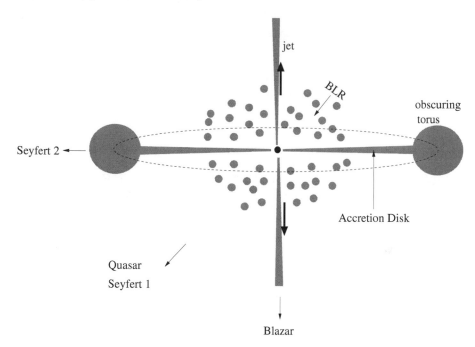

FIGURE 17.6: The cross-sectional view of an AGN. The central black hole is surrounded by an accretion disk and an obscuring torus. The two jets of charged particles are emitted perpendicular to the plane of the accretion disc. The central region is surrounded by the Broad Line Region (BLR), which has high temperature and leads to the formation of broad spectral lines. If this region is in the line of sight of the observer, then the AGN appears as a Quasar or a Seyfert 1 object. The torus obscures this hot medium. Hence if the line of sight is close to the plane of the torus, then we only see narrow spectral lines formed by cooler medium at larger distances from the black hole. The AGN appears as a Seyfert 2 object in this case. Finally it appears as a blazar if viewed directly along one of the jets.

Exercises

17.1 The observed radio flux of Cygnus A is about 31,700 Jy at 14.7 MHz. It is located at a distance of roughly 230 Mpc. Assume that its flux satisfies the power law, Equation 17.9, in the frequency range 10 MHz to 2 GHz, with $\alpha = 0.7$. Compute its radio luminosity in this frequency range assuming that it emits isotropically.

17.2 Determine the upper limit on the size of an AGN whose luminosity shows a significant change in a time interval of about 1 hour.

17.3 Determine the distance of the quasar 3C273, located at $z = 0.158$, using Equation 16.6. Hence determine its absolute visual magnitude, given that $V = 12.8$. Finally determine its visual luminosity, assuming that it radiates isotropically.

17.4 Consider the motion of a segment of an AGN, shown in Figure 17.2. Let the light signal from A and B reach the observer at times t_1 and t_2, respectively. Show that $\Delta t = t_2 - t_1$ is given by Equation 17.5.

17.5 Determine the angle θ for which the apparent speed u in Equation 17.6 is maximum. You can determine this by requiring that $du/d\theta = 0$.

17.6 Following the steps explained in the text, derive the Eddington limit, Equation 17.2, on the luminosity of an object of mass M.

Appendix

Fundamental Constants and Conversion of Units

Gravitational constant $G = 6.67384 \times 10^{-11}$ m^3/(Kg $\cdot$ s^2)

Planck's constant $h = 6.6261 \times 10^{-34}$ Kg $\cdot$ m^2/s

$\hbar = \frac{h}{2\pi}$

Boltzmann constant $k = 1.38 \times 10^{-23}$ Kg m^2/(K $\cdot$ s^2)

Stefan-Boltzmann constant $\sigma = 5.670 \times 10^{-8}$ W/(m$^2 \cdot$ K^4)

speed of light $c = 2.9979 \times 10^8$ m/s

mass of proton $m_p = 1.6726 \times 10^{-27}$ Kg

mass of electron $m_e = 9.1094 \times 10^{-31}$ Kg

Mass of Sun $M_S = 1.9891 \times 10^{30}$ Kg

Mass of Earth $M_E = 5.972 \times 10^{24}$ Kg

Radius of Sun $R_S = 6.95500 \times 10^8$ m

Luminosity of Sun $L_S = 3.846 \times 10^{26}$ Joules/s

1 Joule (J) $= 10^7$ ergs

1 eV $= 1.6022 \times 10^{-19}$ J

1 year $\approx 3.156 \times 10^7$ s

1 parsec $= 3.086 \times 10^{16}$ m

1 A. U. $= 1.496 \times 10^{11}$ m

1 Jansky (Jy) $= 10^{-26}$ W/(m$^2 \cdot$ Hz)

1 barn $= 10^{-28}$ m^2

1 Pascal (Pa) $= 1 \ N/m^2 = 1$ Kg/(m $\cdot$ s^2)

References

1. F. H. Shu, *The Physical Universe, An Introduction to Astronomy*, University Science Books, Sausalito, CA (1982).

2. D. D. Clayton, it Principles of Stellar Evolution and Nucleosynthesis, The University of Chicago Press (1983).

3. H. Karttunen, P. Kröger, H. Oja, M. Poutanen and K. J. Donner, *Fundamental Astronomy*, Springer-Verlag (1984).

4. B. W. Carroll and D. A. Ostlie, *An Introduction to Modern Astrophysics*, Addison-Wesley Publishing Company Inc. (1996).

5. M. A. Seeds, *Stars and Galaxies*, Brooks/Cole Thomson Learning (2001).

6. A. R. Choudhuri, *Astrophysics for Physicists*, Cambridge (2010).

Index

Printed and bound by CPI Group (UK) Ltd, Croydon, CR0 4YY

23/10/2024

01777672-0015